Development of a Remote Laboratory for Engineering Education

Technology Guides: Advancing Capacity Building in Contemporary Organizations

Series Editors:

Hamid Parsaei – Professor of Industrial and Systems Engineering at Texas A&M University (TAMU), Mohamed Y. Ismail – Senior IT Consultant, Texas A&M University at Qatar (TAMUQ), and Boback Parsaei – Senior Engineering Consultant, Integrated Technology Systems, Inc., Texas

This new series, dedicated to the focus book line, is intended to provide a timely coverage of recent technological advancements, with clear and strong focus on breakthrough science and engineering, and the practical utility of such discoveries to present day organizations. Each book in the series will act as a short practical guide for an emerging technology or portfolio of technologies while discussing real-world applications that could assist organizations in responding to customer demands in order to increase market share. The new series should be of value to all practitioners as well as those interested in understanding the basics of new products and processes. The books will appeal to academics, practitioners, and students.

Development of a Remote Laboratory for Engineering Education

Ning Wang, Qianlong Lan, Xuemin Chen, Gangbing Song, and Hamid Parsaei

For more information on this series, please visit: https://www.crcpress.com/%20Technology-Guides-Advancing-Capacity-Building-in-Contemporary-Organizations/book-series/%20CRCTECADVCA

Development of a Remote Laboratory for Engineering Education

Ning Wang, Qianlong Lan, Xuemin Chen,
Gangbing Song, and Hamid Parsaei

CRC Press
Taylor & Francis Group
Boca Raton London New York

CRC Press is an imprint of the
Taylor & Francis Group, an **informa** business

First Edition by CRC Press
6000 Broken Sound Parkway NW, Suite 300
Boca Raton, FL 33487-2742

First issued in paperback 2021

© 2020 Taylor & Francis Group, LLC
CRC Press is an imprint of Taylor & Francis Group, an Informa business

No claim to original U.S. Government works

ISBN 13: 978-1-03-223748-0 (pbk)
ISBN 13: 978-0-367-33441-3 (hbk)

Visit the Taylor & Francis Web site at
http://www.taylorandfrancis.com

and the CRC Press Web site at
http//www.crcpress.com

Library of Congress Cataloging-in-Publication Data

Names: Wang, Ning (Research scientist), author.
Title: Development of a remote laboratory for engineering education / Ning Wang, Qianlong Lan, Xuemin Chen, Gangbing Song, and Hamid Parsaei.
Description: Boca Raton, FL : CRC Press, [2020] | Series: Technology guides. Advancing capacity building in contemporary organizations | Includes bibliographical references and index.
Identifiers: LCCN 2019055183 (print) | LCCN 2019055184 (ebook) | ISBN 9780367334413 (hardback) | ISBN 9780429326455 (ebook)
Subjects: LCSH: Engineering experiment stations. | Engineering—Study and teaching. | Distance education.
Classification: LCC TA416 .W36 2020 (print) | LCC TA416 (ebook) | DDC 620.0071/1—dc23
LC record available at https://lccn.loc.gov/2019055183
LC ebook record available at https://lccn.loc.gov/2019055184

Typeset in Times
by codeMantra

CONTENTS

PREFACE

With the rapid development of Internet of Things (IoT) technology, remote laboratory (RL) technology has been widely used to develop remotely accessible networked equipment for engineering education, scientific research, and industrial applications. As a pillar of online education and scientific research, RL technology has made great progress. An RL system essentially consists of a set of physical experiment equipment linked to the Internet, and users can access and interact with this system through an Internet connection. RL systems eliminate the geographical constraints of conducting experiments, making them a highly effective tool in helping a wide range of users obtain practical experiences needed for competency in science and engineering.

The primary objective of this book is to present a complete guide to the development of an innovative advanced RL technology and address commonly encountered issues among current research and applications in this area. To this end, a new unified framework that focuses on integrating existing offline laboratory experiments under the control of global and local scheduler web servers via Internet is designed and developed in this book. With this novel framework, RL systems can provide an efficient and reliable way for resource sharing in education, scientific research, and industry.

The book is comprised of seven chapters discussed in further detail below. Chapter 1 reviews current RL technology literature. Chapter 2 introduces a new flexible framework for rapid integration of existing offline experiments into RL systems. This framework offers the ability to customize remote experiments for engineering education, research, or industrial applications. Chapter 3 presents an

RLaaS-Frame framework (RLaaS: Remote Laboratory as a Service) based on cloud computing for rapid deployment of RL systems. Chapter 4 provides a novel mobile-optimized RL application architecture. This mobile application architecture has been designed and implemented successfully as a tool for a broad range of fields such as M-learning and industrial electronics to name a few. Chapter 5 introduces an online programmable platform for remote programmable control experiment development. Chapter 6 presents a Wiki-based RL platform for engineering education. It is an innovative approach that has been proposed for RL platform development based on a combination of advantages of Wiki technology and the flexible unified framework. Chapter 7 provides three case studies of successful remote experiments: a remote shape memory alloy (SMA) experiment, a remote proportional–derivative–integral (PID) motor speed control experiment, and a programmable remote robot control experiment. Chapter 8 concludes with possible future research topics.

MATLAB® is a registered trademark of The MathWorks, Inc. For product information,
    please contact:
    The MathWorks, Inc.
    3 Apple Hill Drive
    Natick, MA 01760-2098 USA
    Tel: 508-647-7000
    Fax: 508-647-7001
    E-mail: info@mathworks.com
    Web: www.mathworks.com

AUTHOR BIOGRAPHIES

Ning Wang, Ph.D., is a research scientist in an oil & gas equipment company. He received his B.S. degree in the Information Management System from China Agriculture University (CAU) in 2002, Beijing, China. He received his M.S. degree in Software Science from Hong Kong Polytechnic University (HKPU) in 2008 and the M.S. degree in Computer Science from Texas Southern University (TSU) in 2014. He received his Ph.D. degree in Electrical Engineering from the University of Houston (UH) in 2017. He was a Postdoctoral Research Fellow in the National Science Foundation (NSF) Center for Research on Complex Networks at Texas Southern University (TSU) from 2017 to 2019. He received the College of Science, Engineering and Technology (COSET) Distinguished Graduate Student Award from Texas Southern University in 2014. He has published over 23 peer-reviewed journal papers and 12 conference papers. His research interests include remote laboratory technology and remote sensing, machine learning, Internet of Things (IoT), and Wireless network technology. He is a member of the Institute of Electrical and Electronics Engineers (IEEE).

Qianlong Lan received his B.S. degree in Network Engineering from Shanghai Second Polytechnic University, Shanghai, China, in 2013, and received his M.S. degree in Computer Science from Texas Southern University (TSU), Houston, Texas, USA, in 2016. He is currently pursuing his Ph.D. degree in Electrical Engineering from the University of Houston, Houston, TX, USA. His research interests include MRI RF safety, remote laboratory, remote control, machine learning, artificial intelligence, and big data.

Xuemin Chen received his B.S., M.S., and Ph.D. degrees in Electrical Engineering from the Nanjing University of Science and Technology (NJUST), China, in 1985, 1988, and 1991, respectively. He started his academic career at NJUST and was a faculty member of its automation department from 1991 to 1998. Following this, he was a Postdoctoral Fellow and then a Research Assistant Professor in the Department of Electrical and Computer Engineering at the University of Houston from 1998 to 2006. He was the recipient of the Top Research Innovations and Findings Award from the Texas Department of Transportation (TxDOT) for his contribution in the project "Thickness Measurement of Reinforced Concrete Pavement by Using Ground Penetrating Radar" in 2004. He joined Texas Southern University (TSU) in 2006. Currently, he is a Professor in the engineering department at TSU. His research interests are in virtual and remote laboratory, wireless sensor networks, Internet of Things, and structural health monitoring. He initiated the Virtual and Remote Laboratory (VR-Lab) at TSU and has served as a founding director of VR-Lab since 2008. With the support of the National Science Foundation (NSF) HBCU-UP, CCLI and IEECI programs, and the Qatar NPRP Cycle 4 Award, he has established a state of the art VR-Lab at TSU. He is a senior member of the Institute of Electrical and Electronics Engineers (IEEE).

Gangbing Song received his Ph.D. and M.S. degrees in Mechanical Engineering from the Department of Mechanical Engineering at Columbia University in the City of New York in 1995 and 1991, respectively. He received his B.S. degree in 1989 from Zhejiang University, China. He is the founding Director of the Smart Materials and Structures Laboratory and a Professor of Mechanical Engineering, Civil and Environmental Engineering, and Electrical and Computer Engineering at the University of Houston (UH). He holds the John and Rebeca Moores Professorship at UH. He is a recipient of the National Science Foundation (NSF) Career Award in 2001. He has expertise in smart materials and structures, structural vibration control, piezoceramics, ultrasonic transducers, structural health monitoring, and damage detection. He has developed two new courses in smart materials and published more than 400 papers, including 200 peer-reviewed journal articles. He is also an inventor or co-inventor of 11 US patents and 11 pending patents. He has received research funding in smart materials and related research from NSF, DoE (Department of Energy), NASA, Department of Education, Texas Higher Education Board, TSGC (Texas Space Grant Consortium), UTMB (University of Texas Medical Branch), OSGC (Ohio Space Grant Consortium), OAI (Ohio Aerospace Institute), ODoT (Ohio Department of Transportation), Hewlett-Packard (HP), OptiSolar, General Electric (GE), and Cameron. In addition to his research efforts, he has a passion for improving teaching using technology. He received the prestigious Outstanding Technical Contribution Award from the Aerospace Division of the American Society of Civil Engineers (ASCE), the Excellence in Research & Scholarship Award at Full Professor Level from UH, the Celebrating Excellence Award for Excellence in Education from ISA (International Society of Automation), the Institute of Electrical and Electronics Engineers (IEEE) Educational Activities Board Meritorious Achievement Award in Informal Education, among others. He is a member of ASCE, the American Society of Mechanical Engineers (ASME), and IEEE. He served as the General Chair of the Earth and Space Conference 2010, Aerospace Division, ASCE.

Hamid Parsaei, Ph.D., P.E., is an internationally recognized leader in the field of engineering education, manufacturing systems design, leadership, and economic decision making with applications to advanced manufacturing systems with more than 35 years of experience. He served as Associate Dean for Academic Affairs, Director of Academic Outreach, and Professor of Mechanical Engineering at Texas A&M University at Qatar and held the rank of Professor in the Industrial and Systems Engineering at Texas A&M University in College Station. He also served as Professor and Chair of the Industrial Engineering at the University of Houston for ten years.

He has been the principal and co-principal investigator on projects funded by the National Science Foundation (NSF), Qatar Foundation, the US Department of Homeland Security, National Institute of Standards and Technology (NIST), National Institute for Occupational Safety and Health (NIOSH), Texas DoT (Department of Transportation), among others, with total funding in excess of $26 million. He has authored or co-authored more than 280 refereed publications in archival journals and conference proceedings. He has held several key leadership positions with the Institute of Industrial and Systems Engineers (IISE). He is also a Fellow of the Institute of Industrial and Systems Engineers (IISE), American Society for Engineering Education (ASEE), Society of Manufacturing Engineers (SME), and Industrial Engineering and Operations Management Society International (IEOM),

He is a registered professional engineer in the state of Texas.

INTRODUCTION

In the era of information and Internet technology explosion, especially with the Internet of Thing (IoT) technology's rapid development, remote laboratory (RL) technology has been widely used to develop the remote accessible networked equipment for education, scientific research, and industrial applications. To match the increased requirements of the various fields, a new-generation unified framework is proposed and implemented in this book.

I.1 MOTIVATIONS AND OBJECTIVES

As the important component of online education and scientific research, the RL technology has made great progress. RL systems essentially consist of a physical experimental equipment linked to the Internet, and users can access and interact with them through an Internet connection [1, 2]. As RL systems free experiments from geographical constraints, it can be a highly effective tool in helping a wide range of users regardless of regional boundaries to obtain practical experiences needed for competency in science and engineering. To significantly reduce the cost of maintaining a wide variety of experimental equipment, more and more RL systems are being designed and developed by many research groups in different institutes of the world for educational and industrial applications [3, 4]. However, some challenges still are not resolved in RL systems' design and implementation yet. Currently, there are five essential issues existing in the research of RL technology. These essential issues are listed as follows:

1. How to design a flexible approach to rapidly integrate a new offline experimental equipment into an RL system [5–7]?
2. How to rapidly deploy a stable RL system for widespread usage [8–11]?
3. How to design a mobile RL application that can run cross-platform easily with low cost and less development efforts [12–14]?
4. How to develop an online programmable platform to support users to implement their own designed control algorithm and model to real-time control the real experimental equipment remotely [15–18]?
5. How to develop a well-structured and coordinated online experiment platform to improve engineering education and research activities [19–20]?

The general goal of my research is to develop an innovative advanced RL technology to resolve these five essential issues listed above. To achieve this goal, in this book, the design and development of a new-generation unified framework, which focuses on integrating existing offline laboratory experiments under the control of global and local scheduler web servers via Internet, is discussed. The detail sub-objectives are listed as follows.

First, this research will develop an innovative RL technology with Web 2.0 concepts and HTML5 technology embraced for RL system developers. With this novel framework, the users can easily bring the existing offline laboratory experiments online without extensive knowledge of networking, data transmission, or Internet protocols.

Second, the research will deliver a total solution that is capable of global sharing and distributing RL system resources to the Internet users with cloud computing technology. With this proposed framework, all the available RL systems can be shared among departments, institutions, and countries.

Third, the new RL systems developed with this framework will let the users run the entire RL system from a webpage that does not require plug-ins and is independent from the terminal devices (desktops, laptops, mobile devices, etc.), operating systems (Windows, Linux, iOS, etc.), and web browsers (Microsoft Edge, Chrome, Safari, etc.). The users can go online to conduct experiments without having to worry about which runtime engine version or plug-in is needed. Based on this novel framework, the new RL systems can provide efficient and reliable way for resource sharing in education, scientific research, and industry.

I.2 CONTRIBUTIONS

In this book, as shown in Figure I.1, a new-generation unified framework for RL system development is established. This new-generation unified framework includes five parts, a novel flexible framework for rapid integrating offline experimental equipment [21–24], an RLaaS-Frame framework (RLaaS: Remote Laboratory as a Service) based on cloud computing for RL systems' rapid deployment, a novel mobile-optimized RL application architecture [25], a novel online programmable platform for remote programmable control experiment development, and a Wiki-based RL platform. With this new-generation unified framework [26], these five essential issues in RL system development are addressed very well.

With the flexible framework, this new unified framework provides a flexible approach to rapidly integrate previously offline experiments into RL systems. To rapidly deploy an RL system for widespread use, an RLaaS-Frame framework that leverages cloud computing

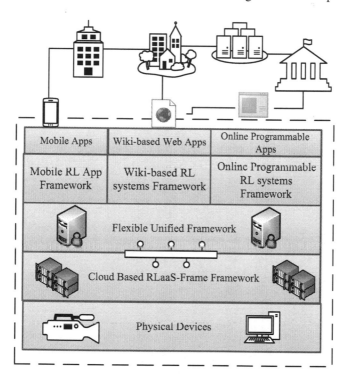

Figure I.1 A new-generation unified framework.

is developed. Meanwhile, an RL application architecture that is optimized for mobile applications is developed to allow easy cross-platform operation while requiring low cost and development effort. Moreover, a new online programmable platform is developed to allow users to implement their own control algorithms and models. Finally, a Wiki-based RL platform is implemented for engineering education. The above components will help guide RL system design and allow for rapid and efficient development in future.

To better demonstrate the effectiveness of the new-generation unified framework, three remote experiments are discussed as the case studies [27–29]. Compared with traditional RL system development approaches, the new-generation unified framework offers a more flexible way to support design and implementation of users' RL systems. It will significantly benefit the RL technology development in future.

I.3 ORGANIZATION

This book is organized into eight chapters. The "Introduction" part introduces the motivations, objectives, and contributions of this book.

Chapter 1 provides reviews of the current literature on technical fields related to RL technology. The system structure, development technologies, and software and hardware platform implementation approaches of RL system are introduced. An introduction to applications of RL technology in the different fields and future potential applications is also included in this chapter.

In Chapter 2, a new flexible framework for rapidly integrating offline experiments into RL systems is presented. This flexible framework has been designed and implemented successfully with a combination of advantages of social online instant messaging (IM) application architecture and the new version Laboratory Virtual Instrument Engineering Workbench (LabVIEW) to Node.js (LtoN) module for experimental equipment real-time control. With the software-plug-in-free real-time communication solution based on the innovative LtoN module, it effectively reduces the efforts to shorten the time of RL system integration. By assigning a unique ID for the existing experiment, it provides a more flexible approach to integrate the existing offline experiment into the RL system. With this new framework, a more powerful online distributed RL system has been delivered. It offers a more flexible way to build up different pattern remote experiments for engineering education, research, and industrial applications.

In Chapter 3, an RLaaS-Frame framework based on cloud computing for RL systems' rapid deployment is presented. Based on the cloud computing technology, the RLaaS-Frame has been successfully established for RL systems' rapid deployment. With the RLaaS-Frame framework, the real-time data communication challenge discovered in the RL system implementation can be solved as well. As a pilot RL platform based on the RLaaS-Frame, a Wiki-based RL platform and the mobile-optimized application architecture have been integrated into the RLaaS-Frame successfully.

In Chapter 4, a novel mobile-optimized RL application architecture is presented. This mobile application architecture has been designed and implemented successfully to provide a new tool for many fields' applications development, such as M-learning application and industrial electronics applications. It integrates the advantages of both native mobile applications and web applications. Meanwhile, it improves the running performance and solves the hardware accessibility issue of web applications. The cross-platform running issue of native application is solved as well. Moreover, it seamlessly combines the unified framework and ionic framework together to deliver excellent RL services to end users.

In Chapter 5, an online programmable platform for remote programmable control experiment development is presented. This new programmable platform has been designed and implemented successfully. To implement this new online platform, an open-source web-based integrated development environment (IDE), Cloud9, is integrated into the Wiki-based RL platform. It offers a new approach to freely control an experiment with a fixed control model and allows users to input their own designed control algorithm code via webpage without any other software plug-ins. In addition, a new online tool is offered for users to promote their active participation in the lab practice as well.

In Chapter 6, a Wiki-based RL platform for engineering education is presented. It is an innovative approach which has been proposed for RL platform development based on a combination of advantages of Wiki technology and the flexible unified framework. With this architecture, a more powerful online learning system supported by remote experiments has been delivered. Additionally, it offers a more flexible way to support students' collaborative learning.

In Chapter 7, the detail design and implementation process of three remote experiments are presented: a remote shape memory alloy (SMA) experiment, a remote proportional–derivative–integral

(PID) motor speed control experiment, and a programmable remote robot control experiment. These three remote experiments based on the new-generation unified framework are used to better guide the development of RL systems in future.

Chapter 8 closes the book with conclusions related to the research topics discussed in the previous chapters. Possible future research topics related to the work presented in the book are provided as well.

1

INTRODUCTION OF REMOTE LABORATORY TECHNOLOGY

1.1 INTRODUCTION

Laboratory tests and experimentation are essential components for education and scientific research in all fields of engineering [30–32]. Hands-on laboratories are the most common forms to provide users with opportunities of experimentation in physical, or real, systems related to the scientific research, education, or industrial applications. In spite of offering the actual experience in scientific research, education, or industrial application, hands-on laboratories are known for their high costs associated with the required equipment, space, and maintenance staff [33, 34]. With the significant technological advancements of computer and network technology, remote laboratory (RL) technologies have been introduced into education, scientific research, and industrial electrical applications, and also integrated into e-learning frameworks offered to engineering and science students [35, 36]. It not only personalizes the learning pathways, but also reduces the costs significantly and makes laboratory experiments available almost anytime and anywhere [19, 20]. RL, by definition, is a laboratory that enables the users to conduct an experiment remotely through the Internet, and its accessibility through the Internet can resolve cost and access constraints as it can be used at flexible times and from various locations [19, 20, 37]. Compared to traditional hands-on experiments, experiments operated remotely over the Internet offer many advantages. One of these benefits is the capability to handle a large number of users to conduct experiments through scheduling. The RL allows a workaround for complex logistics, such as staff, space, scheduling, budget, and commuting. The RL systems can always be made available to users who may conduct the experiments from any Internet accessible device, such as a desktop, laptop, and mobile device. RL systems could set the stage for shared infrastructure between departments and colleges, as well as amongst different institutions around the world [38]. The capability to share real experiments makes experiments significantly more accessible and affordable.

RL technology is not a new concept nowadays, and it has been around since the introduction of the Internet in the 1970s [39, 40]. It has slowly evolved from the union of online-learning programs, such as MIT Open-Course-Ware and PROLEARN [41]. RL systems essentially consist of a physical experiment linked to the Internet, and they can be accessed by users who can also interact with them through an Internet connection. As an RL system enables experiments to be free from geographical constraints, it can be a highly effective tool in helping a wide range of users regardless of regional boundaries to obtain practical experiences needed for improving their competency in education, scientific research, and industrial applications. This location-independent access is especially useful in situations where space is limited, as well as in the distance applications in education, scientific research, and industry [42]. Compared to traditional hands-on experiments, RL systems can add more values to the experiments conducted in the following ways:

- They can offer similar flexibility benefits as simulation-based experiments, when laboratory space and students' schedules are considered.
- Unlike conventional lab sessions, access to remote experiments can be 24 hours, 7 days a week.
- Users can use RL systems as a supplement or replacement for traditional laboratory assignments.
- Overall better scheduling of activities.
- Better return on investments in equipment due to shared resources.
- Enable education and research collaborations between individuals and institutions all around the world.
- Enable and support autonomous learning.
- Prevent damages to equipment via the integration of practice sessions on virtual labs prior to experimentation, and ensure user safety at all times.
- Meet the experimentation needs of distance applications in education, scientific research, and industry.

1.2 CURRENT STATUS OF THE RL TECHNOLOGY

To date, more than 250 RL systems for different types of applications can be found around the world as a result of both individual and large-scale, multi-national, multi-institutional efforts [19, 20]. A large variety of experimental configurations are available; newer modalities of RLs are beginning to appear, including the triple-access mode (hands-on, virtual, and remote) laboratory based on Laboratory Virtual Instrument Engineering Workbench (LabVIEW) [43] and even multi-user RL systems [44]. Some of the most prominent examples of RLs are briefly described with emphases on their purpose, architecture, and platform in Table 1.1.

TABLE 1.1
Examples of RL Systems

RL System Examples	Institution	Description from Vendor
iCampus iLabs [45–47]	Massachusetts Institute of Technology (MIT)	The MIT iLab has developed a "distributed software toolkit and middleware service infrastructure to support Internet accessible laboratories and promote their sharing among schools and universities on a worldwide scale." The iLab is an RL based on client–server architecture.
WebLab-Deusto [48, 49]	University of Deusto	WebLab-Deusto is an "open-source distributed remote lab used with students at the University of Deusto since February 2005 as an essential tool for their practice work in different engineering-related subjects It makes it possible to offer real experiments (e.g: FPGA, CPLD, PIC microcontrollers...) to a certain group of users through any computer network." This RL uses browser–server architecture software.

(Continued)

TABLE 1.1 (*Continued*)
Examples of RL Systems

RL System Examples	Institution	Description from Vendor
Virtual Instrument Systems in Reality (VISIR) [50–52]	Blekinge Institute of Technology, University of South Australia	The VISIR has "online laboratory workbenches for electrical experiments that mimic traditional ones by combining virtual and physical reality." The VISIR uses client–server architecture software.
UTS Remote Lab [53, 54]	University of Technology Sydney	UTS Remote Lab is "part of Lab Share, an Australian Government funded project that aims to create a national network of shared, remotely accessible laboratories." This lab also uses client–server architecture software, and an RL consists of experimental apparatus.
LiLa (Library of Labs) [55, 56, 57]	Eight European universities and three enterprises	The goal of LiLa is "the composition and dissemination of a European infrastructure for mutual exchange of experimental setups and simulations, specifically targeted at undergraduate studies in engineering and science." In LiLa, there are virtual laboratories (simulation environments) and RLs (real laboratories that are remotely controlled via the Internet). For the implementation of RLs, a browser plug-in can be used to gain control over all devices. LiLa thus uses a browser–server architecture RL system with plug-in capability.
eComLab [37]	The University of Texas at San Antonio	The eComLab system uses the Relay Gateway Server (RGS) architecture to form an RL consisting of experimental apparatus that can be remotely monitored and controlled via the Internet.

(*Continued*)

TABLE 1.1 (*Continued*)
Examples of RL Systems

RL System Examples	Institution	Description from Vendor
Smart Material Remote Labs [58]	University of Houston (UH)	The goal of the RLs at UH is to provide users with an interface that will work in most Internet-enabled web browsers without the need to install most of the software plug-ins.
VR-Lab [59]	Texas Southern University (TSU)	TSU collaborated with UH to develop a plug-in-free remote experimental platform. Virtual and remote data communication experiments and the remote Digital Signal Processing (DSP) experiment were developed. Their RL is also a browser–server architecture software system.
RL at TAMUQ [21]	Texas A&M University at Qatar (TAMUQ)	The RL at TAMUQ was established through the collaborative effort of TAMUQ, University of Houston, and Texas Southern University. Two experiments, the smart vibration platform (SVP) experiment and the shape memory alloy (SMA) experiment, form the initial collection of this laboratory. A novel unified framework was used to implement the RL platform.

For the development of these RL systems, there are mainly two kinds of system architectures: client–server architecture and browser–server architecture. In the early stages of RL development, client–server architecture software system was used for achieving high-performance, real-time, experimental data transmission [60, 61]. One of the popularly deployed technologies at that time for remote panel over the Internet was National Instruments (NI) LabVIEW [62, 63]. Currently, with client–server architecture, some other newly integrated learning environments for remote experimentation are being developed based on web and .NET remote services. These new learning environments are being deployed for students working in different locations to simultaneously and collaboratively complete complex experimental exercises [64].

With the development of computer technology, the browser–server architecture is becoming more stable. Especially, the REST (Representational State Transfer) provides a lightweight protocol accessible to a wide variety of clients. To adopt the browser–server architecture, LabVIEW integrates a new feature to interact with the Virtual Instruments by using the RESTful web services. This architecture does not require complex message passing, but rather it provides a simple interface for the user to use web services in LabVIEW. However, it requires the client interface to be developed using different technologies, such as LabVIEW plug-in and Flash plug-in [58]. However, as the number of remote experiments increases, capacity of the software to handle multiple users with multiple resources reduces.

As shown above, RL technology is a vibrant area of research and continuous development which leads to finding new and improved ways to deliver laboratory experience across distances. However, it is also equally important to note that to date each RL has been a stand-alone application. Currently, most of the existing RL systems have their own Learning Management System (LMS) based on web service technology, and they use software plug-ins to handle the real-time communication between client web application and experimental equipment. For example, if one of the institutions listed in Table 1.1 plans to share a particular laboratory with another institution, it is very difficult due to fundamental differences between their infrastructures, service providers, platforms, programming languages, etc. Their inability to share resources is the single greatest drawback of current RL system solutions in terms of large-scale implementation to reach an exponentially larger number of users.

1.3 THE DEVELOPMENT OF RL SYSTEMS

The technologies used in the development of RL systems are various. In the past decades, NI LabVIEW and MATLAB®/Simulink® were the major software tools used for the development of RLs' experimental environment. The server majorly uses Apache web engine in Linux operating system. MySQL database system has been the most widely used for experimental data database development since the late 1990s. Java, HTML, JavaScript, PHP (Hypertext Preprocessor), and Adobe Flash are all popular choices for the development of graphical user interface (GUI). Majority of real-time videos are played by the

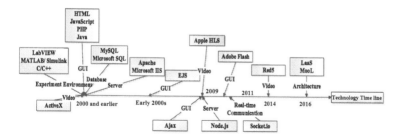

Figure 1.1   Time line of technology development in RL systems.

ActiveX components embed in the web pages. Figure 1.1 depicts a time line of technology development in RL based on public information and references [45, 53, 57, 65–70].

Most major software and technologies found in recent laboratories have been in use from 2000 or earlier. According to literature and public information, there are at least four new technologies (the results may not represent all RLs) found in recent RLs. In 2004, Easy Java Simulations (EJS) was released. In 2005, an RL presented in [66] used EJS for the development of GUI. In 2006, Asynchronous JavaScript and XML (Ajax) was released. In 2008, iLab by MIT successfully adopted Ajax in the development of GUI [45]. In 2009, Adobe Flash was released and adopted for use in GUI in RL [66]. Red 5 is an open-source media streaming server whose Version 1.0 was released in December 2012. Two years later, Red 5 was used for real-time video in RL [37].

Most of the early-stage RLs relied on the client–server architecture for achieving high-performance, real-time, experimental data transmission [61, 71]. Later on, a client–server architecture based on web and .NET remote services was developed and deployed for RLs [64]. With continuing improvements in computer performance, the technology supporting the browser–server architecture is becoming increasingly more stable and suitable for cross-platform system design. The importance of adopting cutting-edge web-related technologies, Web 2.0 and web services/service-oriented architecture (SOA) to produce better RLs is addressed. Meanwhile, a large number of new technologies (such as Java applets, Adobe Flash, AJAX, HTML, and ActiveX) have been developed to support more complex web-browser-based Internet applications. Almost all of the current RLs use a similar architecture: the Web 2.0 technology for client web application and the NI LabVIEW or MATLAB for experimental equipment control application, and server software including LMS is normally installed on the server as middleware. Web service technology, including the Simple Object Access Protocol (SOAP), Asynchronous JavaScript and XML (AJAX), and Representational State Transfer (RESTful), is the most common selection for the middleware implementation [65, 45]. The drawback of the web service technology for real-time data communication is its low transmission efficiency [64]. To improve this drawback, some extra plug-ins, such as Java Applet, Adobe Flash, MATLAB Simulink, and LabVIEW, are needed.

As discussed above, the common software architecture is composed of the following: the experimental equipment itself; a local controlled workstation connected to the device, which plays the role of a gateway between the experimental devices; and the remote computer of the users.

1.3.1 Cloud Computing Technology for RL Systems' Implementation

Nowadays, the inability to share resources is the single greatest drawback of current RL system solutions in terms of large-scale implementation to reach an exponentially large number of users [8–11]. The cloud computing technology provides a viable solution to address this essential issue. With the concepts and infrastructures for development of cloud technology, several definitions of cloud technology have emerged. The National Institute of Standards and Technology (NIST) provided a more comprehensive description, identifying cloud technology with the parameters and cloud computing architecture [72]. The SaaS (Software as a Service) is the first layer of standard cloud, the PaaS (Platform as a Service) is the second layer of standard cloud, and the IaaS (Infrastructure as a Service) is the third layer of standard cloud. These three layers provide three different series of services to different users.

Cloud computing technology [73] works through several service models, each of which provides differing levels of access and capabilities to the end user. At the base are physical devices such as servers, which connect to the network through a series of access switches forming the access layer. The switches in turn connect to aggregation switches, which provide various higher-level organizational functions, such as domain services and load balancing. These aggregation switches form the aggregation layer. At the top is the core layer, which consists of routers that provide access through the

Internet. Each layer can possess a high level of redundancy to minimize chances of failure [74].

Several other properties of cloud technologies include uniform high capacity and fault tolerance [72]. For the focus of the proposed research, two important properties stand out: fluid adaptation for virtual machines (VMs) and scalability. VMs are specific portions of the cloud infrastructure (e.g., processors, memory) selectively chosen and used to create the sense of a standalone computer. Ideally, the infrastructure can be freely reassigned to other components within the cloud network without the user noticing. This fluid transition allows for further robustness and higher bandwidth for the whole network [74]. This segregation of resources to form VMs and their fluidity can be useful for control of experiments and data storage, such as for RLs that will be discussed later. Another vital component in the proposed research is scalability. For cloud networks, scalability refers to the ability of the architecture to accommodate increasing amounts of servers [74]. In regard to RLs, this refers to the ability of the cloud to easily scale to include more RL modules, which will require the design of a standard to be discussed in the following sections.

Cloud technology is ubiquitous, and the end user in recent times may have experienced the benefits of cloud technology without knowing anything about it. Some of the more prominent examples include Amazon, Microsoft, and Google.

The Amazon Elastic Computing Cloud (EC2) is a well-known example of what cloud technology has accomplished and the flexibility that cloud computing allows at the platform and infrastructure level. The Amazon EC2 provides individuals and organizations a highly inexpensive and elastic access to computational infrastructure. End users are able to reconfigure CPU, memory, and bandwidth through customized web service application programming interfaces (APIs) on demand, as well as launch and manage multiple customizable VMs. Furthermore, VMs in EC2 are run by the Xen virtualization engine [75] and can be placed in geographically disparate regions to avoid potential local problems [76].

The Microsoft Azure Cloud platform provides similar services to the Amazon EC2, albeit with stronger focus on Windows applications and platform-oriented services. Users are able to develop a wide variety of applications (through .Net, Node.js, Java, etc.) on the cloud and automatically scale to fit computing needs. Microsoft Azure is hybrid ready, thus providing ease of integration with the end user's existing infrastructure. Like Amazon's EC2, the physical components are dispersed geographically, even acting as the first international cloud provider in mainland China. A large number of prominent businesses, such as 3M and Xerox, have chosen Microsoft Azure as their cloud solution. Commonly used applications including Skype, Office 365, and Bing are all powered by this platform [77].

The Google App Engine is a more dedicated PaaS cloud solution, allowing users to build applications on Google's infrastructure. However, instead of general-purpose applications that can be supported by EC2 and Azure, the Google App Engine specializes for the support (e.g., provision of Software Development Kits (SDKs)) of web applications, especially those built with Python, Java, PHP, Go, and MySQL [78, 79].

A recent extension to cloud technology is mobile cloud computing (MCC). Mobile devices are commonplace and have become an essential integration of life. In addition to increased battery life, storage, and bandwidth, evolving mobile computing power has allowed ever more powerful applications to be used on mobile devices. By transferring computation and storage to a cloud and creating MCC, mobile computing has and will become even more powerful [80]. The architecture of MCC is similar to regular cloud-based technologies. Local mobile networks consist of access points, and even satellites interface with requests from mobile devices and send the computational requests to a cloud service provider [81, 82]. It is expected that MCC will play an increasingly important role in the increasing presence of cloud technology and also for RLs.

In regard to more academically oriented applications, the iLab Solutions initiated by the Baylor College of Medicine is a leading example of how educational and research applications can be unified through the SaaS model. The iLab Solutions offers a specialized collection of laboratory management software with functions that span from service request management, billing, and invoicing to equipment reservations. Such services are streamlined through the iLab Solutions cloud; for instance, users are able to easily search for any equipment or services that are logged into the system to check for availability and submit a request. As the iLab Solutions is at the SaaS level, all of the infrastructure and platform management is hosted and maintained by the iLab Solutions group and outside the view of the user. The user only needs to set up an account in order to start accessing the software services. The iLab Solutions now has an international coverage of over 10 countries across more than 100 research institutions, 30 of which are the top recipients of National Institutes of

Health funding [83]. From the perspective of the proposed research, iLab Solutions provides the closest form of service model that will be used for RLs in science and engineering education.

Several research activities that are currently carried out to develop a suitable cloud-based solution for a flexible and effective existing RL system are summarized in the following. A research team proposed the Laboratory as a Service (LaaS) model for developing and implementing an RL system at the Spanish University for Distance Education (UNED) [8, 84]. They put forward a general cloud-based RL architecture for distributed LMS and used Java plug-in and Java virtual machine (JVM) to tackle the real-time data transmission issue. However, two essential issues still cannot be solved are as follows: (1) there is no standard definition of the LaaS model for other users, and it is hard to be used for rapid and flexible deployment of future RL; and (2) there is still a plug-in issue for the web RL application to support real-time data transmission. A cloud-based remote virtual prototyping platform was developed at Karlsruhe Institute of Technology (KIT) [10]. They presented a cloud-based architecture allowing the design of virtual platforms and prototyping of the system including sophisticated software with prerecorded data or test benches. However, this architecture is not a standard definition of cloud computing model. It is only a special cloud-based platform for the design and test of multi-processor systems on chips instead, which makes it hard for efficient and flexible deployment of RLs in future as well. As a special cloud-based platform, it has a restricted solution to support real-time data transmission. Moreover, to expand the scope of RL system application into online education, the concept of Massive Open Online Labs (MOOLs) was presented in Longo et al.'s [85] and Salzmann et al.'s [86] papers. Their approach is to deploy RLs within Massive Open Online Course (MOOC) software systems and infrastructures.

1.3.2 Mobile RL Applications Development

To offer students a more flexible way to access RL, instead of forcing the learners to sit in front of a fixed computer to use a location-independent environment for experimentation, a technology that is suitable for presenting RL on mobile devices becomes essential. In addition, integrating the RL into the mobile devices can offer students more flexible approaches to learn and produce better outcomes, as pointed out by May et al. [87] and Silva et al. [88]. So far, most of the current research interests in mobile learning have mainly focused on the various learning theories, and only a smaller number of researches have focused on the design of framework and the mobile device technologies that are compatible with M-learning systems [89, 90]. How to design and implement a mobile-optimized and easy-to-use application for M-learning has become an emerging research topic [91, 92]. Generally, two approaches, the native mobile application and the web-based application, are used to integrate the RL into mobile devices.

- Native RL application for mobile devices:
  Native RL applications are developed using different native codes, different tools, build systems, APIs, and mobile devices with different capabilities for different platforms such as Apple iOS, Android, and Windows Mobile. Meanwhile, native RL applications are compiled, and they directly call the underlying APIs of the different platforms. Although the native RL applications can achieve the best performance on mobile devices with different platforms, it is hard to implement the cross-platform interface [12]. In Table 1.2, examples 1–6 are the native mobile RL applications or mobile industrial applications that were developed with native application approach for different mobile systems. However, they are hard to be migrated to other mobile systems [13].
- Web-based RL application for mobile devices:
  Web-based RL applications are normally created in HyperText Markup Language (HTML), Cascading Style Sheets (CSS), and JavaScript, and they run in different web browsers (such as Safari, Chrome, Microsoft IE, and Firefox) [14].

  In Table 1.2, examples 7–12 are the web-based mobile RL applications. Example 8 used jQuery mobile framework to implement the mobile RL application. Examples 9–12 were implemented by web development technology as well. All of these mobile RL applications can be easily run on different mobile phone systems via different web browsers, while their performance depends on JavaScript rendering and mobile web browsers. With the rapid development of web technology, the web application's running performance has been significantly improved. However, for some special user interfaces (UIs), e.g., example 6 in Table 1.2, the web application's running performance is still a major issue. In addition, some mobile applications need to use sensors in mobile devices, e.g., examples 2 and 5 in Table 1.2. Thus, native application is the only option.

TABLE 1.2

Examples of Mobile RL Applications

Mobile RL Example	Institution	Year	Mobile Application Type	Description from Vendor
1. An Android-based RL application [93]	MIT	2012	Native Mobile Application	Develop a remote Android client application based on MIT iLab shared architecture.
2. A field fluorometer experiment using smartphone [94]	University of Sydney	2015	Native Mobile Application	Develop a novel portable fluorometer using Android-based application framework.
3. J2ME-based mobile laboratory [95]	Princess Sumaya University	2008	Native Mobile Application	Develop a mobile RL application based on J2ME framework.
4. VR-Lab [96]	National University of Distance Education (UNED)	2013	Native Mobile Application	Develop an RL application to support portable devices based on EIS.
5. OMF Measurement Laboratory [97]	University of Malaga	2016	Native Mobile Application	Add a measurement and monitoring tool TestelDroid into Android devices to control remote experiments.
6. 3-D mobile application for remote experiments [98]	Indiana University Northwest	2016	Native Mobile Application	Develop a 3-D application for RL experiments based on 3-D mobile Augmented Reality Interface technology.

(Continued)

TABLE 1.2 (*Continued*)
Examples of Mobile RL Applications

Mobile RL Example	Institution	Year	Mobile Application Type	Description from Vendor
7. A web-based mobile application based on WebLab-Deusto architecture [99]	University of Deusto	2008	Web Mobile Application	Develop an RL using AJAX technology based on WebLab-Deusto architecture.
8. A mobile-accessed RL application for VISIR [50]	Blekinge Institute of Technology (BTH)	2006	Web Mobile Application	Develop an RL using HTML 5 technology and based on WebLab-Deusto platform.
9. Pe͏TEX platform [100]	TU Dortmund University	2011	Web Mobile Application	Develop an RL using HTML 5 technology and based on Moodle LMS.
10. RExLab [101]	Federal University of Santa Catarina (UFSC)	2016	Web Mobile Application	Develop mobile remote experimental application using HTML5 and jQuery Mobile framework.
11. Remote web-based mobile control laboratory [102]	Complutense University	2015	Web Mobile Application	Develop a mobile RL application using HTML 5 technology and Node.js server.
12. iSES Remote Lab SDK for smartphone [103]	University of Prague	2016	Web Mobile Application	Use "iSES Remote Lab SDK" for Arduino-UNO and web technology to develop a mobile RL application.

The other main drawback of web application is the limited access to mobile device hardware.

Currently, most of mobile RL applications are implemented using native application approach and web-based application approach. Native mobile applications are developed for one platform, and they can take full advantage of mobile device capabilities. Web-based applications are not exactly mobile applications but are websites that are mobile formalized. With the continuous improvement of mobile application development technology, developers are migrating to mobile-optimized application development tools such as Phone Gap, jQuery Mobile, Adobe Air, and Titanium to reduce the cost of development and reach out to maximum users across several platforms. A mobile-optimized application is a mix of native and web technologies that are leveraged to deliver a mix of web content and native capabilities. Consequently, the mobile-optimized application provides a new approach for the development of the mobile RL application.

1.3.3 Social Computing Technology for RL System Implementation

As scholars and students are the center of any education and research service, online education and research platform offerings must provide a unique value to maintain high-quality services instead of simply repackaging the content designed for traditional education and research environments [104, 105]. Social computing technology provides an innovative approach for this goal. "Social computing" is a general term in the field of computer science that is concerned with the intersection of social behavior and computational systems [106–108]. Though the term is used to cover a wide collection of implementations and behaviors, there are two broad definitions of social computing as follows:

1. In the weaker sense of the term, social computing is the support of social behavior through computational systems. In this usage, social computing creates or re-creates social conventions and social contexts through the use of software and technology [106, 108]. Social network services, blogs, emails, instant messaging, Wiki, and social bookmarking are few examples of social software that users interact through to conduct social computing.

2. In the stronger sense of the term, social computing has to do with supporting "computations" that are carried out by groups of people [106, 108]. Online auctions, collaborative filtering, user actions analysis and direct push info, prediction markets, reputation systems, computational social choice, tagging, and verification games are all examples of this stronger usage, which emphasizes the use of technology as a platform to allow contributions of many users to a computational task.

Amongst the many common implementations of social computing, Wiki technology is a powerful tool that is useful in presenting collaboratively developed material in a customizable and easily approachable manner [109]. Wiki is a browser–server architecture software system used to facilitate collaborative tasks. Its common uses include group communication, intranets, documentation management and sharing, and publishing. In this system, users interact through Wiki software, a type of collaborative software [110]. Often a website runs a Wiki engine, which allows its users to edit its content and invites all users to edit any pages or to create new pages within the Wiki. A characteristic of Wiki software is that operations are conducted via only web browser by a simplified markup language or a rich-text editor without any extra plug-ins, which makes user modifications without intensive programming or scripting skills possible. Wiki software's basic style looks like a web publishing tool (such as Wikipedia), but it can also serve many different purposes such as knowledge management, note taking, editing and maintenance of materials, and interactive discussion.

Another key aspect of Wiki is the ease of meaningful topic association between different pages in the database through intuitively easy link creation and simple indications of the existence of desired topics in the database. As such, carefully crafted Wikis are not static deposits of information, but rather living documents which invite users to make useful edits and which can easily connect related ideas, leading to a constantly changing and immediately interactive website landscape. Wikipedia, the most famous and popular example of Wiki implementation, illustrates this capability through its nearly endless network of related topics, disambiguation, and external references.

Social computing encompasses a vast array of tools and technologies that carry significant potential for promoting student learning [111]. The gradual integration of social computing into everyday life, especially for those in the younger generation, has generated a new paradigm of learning molded by the benefits provided by social computing. Current and future students increasingly expect social computing capacities in their learning resources; fortunately, experiments in social computing technology and tools show that social computing tools have significant potential for enhancing the learning gains of current students [112, 113]. Some of the key benefits of social computing in helping students learn include (1) supply and distribution of learning materials, (2) networking amongst teachers and students, (3) improved personal and collective knowledge management and resources, (4) access to subject-specific methods and tools, and (5) realization of self-directed learning skills, all of which serve to empower the learners [112].

Technical fields of higher education, particularly engineering, have a particular need for learning tools to help students direct their learning in a way that best fits their needs. Studies in engineering education have shown that student retention in engineering programs is amongst the lowest ones across common university subjects. The reasons students indicate for leaving engineering programs include the steep learning curve of the material and the discrepancy between traditional engineering teaching (lecture based) and active student learning styles (experiential, visual, etc.). Felder et al. [114] showed that many students were visual, experiential, or kinesthetic learners, and those learning types make learning in lecture courses difficult [115]. Students who begin to fall behind on material feel left-behind by the professor in lecture and are often too shy to approach the faculty for individual help. As a result, these students only lag further behind, which is a fact that may not become apparent until test grades come in. This can be remedied by integrating an intensive, low-risk student learning progress feed back system that allows the faculty to easily identify a particular student's areas of strength and weakness.

As it was pointed out in references [5, 6, 19], most of the improved RL solutions mainly focus on the technology innovation, such as software and hardware platform upgrade or use of new IT technology. Developing a well-structured and coordinated RL system to improve engineering education and scientific research has already become an essential issue [20]. Social computing technology provides a suitable approach to address this essential.

2

A NOVEL FLEXIBLE FRAMEWORK FOR RAPIDLY INTEGRATING OFFLINE EXPERIMENT INTO REMOTE LABORATORY SYSTEM

2.1 INTRODUCTION

As discussed in Chapter 1, remote laboratory (RL) technology has been used to develop more and more remote accessible networked equipment for education, scientific research, and industrial applications due to its potency and cost-effectiveness. There are **two issues** that need to be addressed in integrating a remote experiment into an RL system [5–7]: **(1) the employment of special software plug-ins is seriously affecting the cross-platform capability of the RL system, and (2) the special system architecture needs to be designed and implemented to connect the client applications and the experimental equipment.** Current research activities to tackle these issues are summarized in the following.

A unified framework has been developed at Texas Southern University and University of Houston based on the traditional peer-to-peer (P2P) RL system architecture [116]. This unified framework addresses the successful employment of special software plug-ins. However, how to quickly integrate experiment into the RL system is not well addressed by this framework.

A real-time remote access laboratory with distributed and modular architecture was designed at the University of Southern Queensland [117]. They proposed a general architecture for distributed P2P network control systems but only focused on microcontroller unit (MCU) and communication protocol selection. However, how to develop a flexible software system to quickly develop the RL system is still not addressed clearly.

A flexible and configurable architecture for automatic control RLs was designed and implemented at Slovak University of Technology (STU) [118]. They proposed a flexible architecture built on programmable logic controllers (PLCs) and industrial network routers (INRs). However, for the existing experiments not using the PLCs, the new interfaces for experimental hardware and INRs need to be redesigned and reimplemented.

A flexible RL architecture was designed and implemented at the University of Limoges and University of Mostaganem [119]. They proposed a flexible architecture based on Node.js web engine and used a FEHI (flexible Ethernet hardware interface) and PEB (practical evaluation board) to directly handle real-time communication between experimental equipment and Node.js server without using any software tools, such as National Instruments (NI) LabVIEW and MATLAB. However, for an existing experiment, the new hardware interface and PEB need to be reconfigured and reimplemented.

With the quick development of RL technology, the concept of Massive Open Online Labs (MOOLs) was proposed in Lowe's [120] and Salzmann et al.'s [86] papers. Their approach is to deploy RLs within Massive Open Online Course (MOOC) software systems and infrastructures. Due to the limitations in the experimental hardware and the network model as discussed above, it is hard to extend their proposed solutions to other institutes. Therefore, **how to design a flexible approach to rapidly integrate remote experiment into an RL system is still a vital research topic for RL development**. To address this essential issue, a new flexible framework based on social instant messaging (IM) application architecture as a TURN-KEY remote experimental integration solution is proposed and implemented in this chapter. With this new framework, a new remote experiment can be added into the RL system as a node with its unique ID. By using the HTML5 and Web 2.0 technology, the new framework can be easily integrated into most of the popular web-based Learning Management Systems (LMSs), such as Moodle and iLab Shared Architecture (LSA), to connect them to a social network. To the best knowledge of the authors, this solution is the first one using the popular social IM application architecture to integrate the RL system without extra plug-ins. Moreover, a new-version LtoN (LabVIEW to Node.js) module built on Socket.IO protocol is designed and implemented for rapidly connecting the experimental equipment with server.

2.2 METHODOLOGY

The distributed RL system aims to expand to a one-to-many communication paradigm, where one central system serves multiple users, or to a many-to-many (M2M) communication paradigm, with many users using many experiments by different providers [121, 122]. In a distributed system, experimental modules need to be created and hosted by individuals [123, 124]. Users are all scattered in the network, and anyone can connect to any experiment if it is available. To achieve this goal, there are some existing examples for the distributed RL system integration [117–119], and almost all of them focused on developing a general solution to address the essential issue which is, "How to flexibly and rapidly integrate an RL system?" In general, an excellent distributed RL system architecture should have a flexible system architecture, a stable real-time middleware in server, and a high-performance real-time data transmission protocol [117, 118]. Based on our research, online social IM application architecture, Node.js, and Socket.IO are the suitable candidates for implementing a new flexible framework that fulfills the requirements of an excellent distributed system.

2.2.1 Social IM Application Architecture

Social IM application can refer to any kind of real-time communication between any senders and any receivers over the Internet. It addresses one-to-one communications as well as multicast communications (M2M) between many senders and many receivers, or may be a feature of a web conferencing service [125, 126]. For education, the web-based online classroom, which is a kind of social IM application, could provide effective interactive tools and contextual learning scene, and it can deliver two part services, instructional communicating service and collaborative learning environment service [127, 128]. The most notable strength of online IM application is its flexibility and ability to rapidly connect different users together [127]. With this advantage, a new flexible framework is designed and implemented for integrating RLs rapidly and flexibly.

A social IM application has two parts: chat server module and client web module. In the server module, normally, there is a user management module to create the unique user ID for every uscr. Meanwhile, the server module creates a thread based on the unique user ID to support the users' real-time communication. The user management module supports three chat patterns: one to one, one to many, and many to many. The different chat patterns can be flexibly combined based on the users' requirement. The new flexible framework inherits this advantage from the online IM application to provide the three different communication patterns for RL integration. Thus, the new framework offers a more flexible way to build up different pattern remote experiments for engineering education, research, and industrial applications.

2.2.2 Node.js Web Engine

To implement an efficient middleware for supporting high-performance real-time data transmission between experimental hardware and end users, a stable web engine must be chosen. With the development of the Internet of Things (IoT) technology, Node.js, which is known as its speed, scalability, and efficiency, plays the key role with the capability of the reliant on the implementation of the data-intensive applications and the real-time applications [129–131]. With these advantages, it has become a prime candidate for implementation of RL system. Node.js is a web engine that works in the server side, and is designed to notably set up web server for writing scalable Internet real-time communication applications [132]. Node.js uses an event-driven operation mode, asynchronous I/O port to minimize overhead and maximize scalability. The original goal of Node.js is to create web sites with push capabilities as seen in web applications like Gmail [133]. Nevertheless unlike most other JavaScript programs, Node.js is not executed in a web browser. Instead, it is executed as a server-side JavaScript application. In the server, Node.js not only implements multiple common JavaScript specifications but also provides a Read Eval Print Loop (REPL) environment for interactive testing [134]. Comparing with Apache web engine, Node.js is an especially fast and efficient web engine which is more suitable to handle the high-performance real-time data transmission.

Node.js only exposes non-blocking asynchronous interfaces to the programmer. Its power lies in the fact that it stays away from certain undesirable interfaces, such as synchronous I/O. As each web application running on the Node.js is a single thread, the users don't need to consider an event completing and taking over while they are in the middle of another task. Node.js uses the module architecture to simplify the creation of complex web applications [135]. Each module contains a set of functions related to the "subject" of the module.

For example, the Node HTTP proxy module contains functions specific to HTTP Proxy. Furthermore, it provides some core modules to support the user to access files on the file system, to create HTTP Proxy and Socket.IO, and to perform other useful functions. Node.js is also a promising technology and an excellent choice for high-load web applications [136].

2.2.3 Data Transmission Protocol Selection

Currently, most of the RL systems have used the web service technology to handle real-time data transmission. However, the drawback of the web service for real-time data communication is the low transmission efficiency [6], and normally some extra plug-ins, such as Java Applet and Adobe Flash component, are required to fill this gap. With the advent of HTML5, WebSocket protocol provides a new approach to address the low transmission efficiency and extra plug-in issues of web service technology [137]. The WebSocket protocol makes more interaction between a web browser and a server-side middleware possible and facilitates the real-time duplex data transmission between client-side web apps and the server-side middleware. In addition, the real-time data communications can traverse the TCP port 80 with the WebSocket protocol, which is of benefit for those environments blocking non-web Internet connections by using a firewall [138, 139]. Generally, WebSocket is mainly used to solve several key issues that cause the low transmission efficiency with REST and HTTP. These issues include the following: (1) *Bi-directional*: HTTP is a uni-directional protocol. Normally, the client always initiates a request, and the server processes it and returns a response. At last, the client consumes the response from server. However, WebSocket is a bi-directional protocol, and it has no pre-defined message patterns such as request/response [139]. With WebSocket, either the client or the server can send a message to the other part. (2) *Full-duplex*: HTTP allows a request message from client to server, and then the server sends a response message to the client. However, WebSocket allows the client and the server to communicate independent of each other [139]. (3) *Single TCP Connection*: Typically, a new TCP connection is initiated for a HTTP request and is terminated after the response is received. The other new TCP connection should be established for another HTTP request/response. However, for WebSocket, the HTTP connection is upgraded with the standard HTTP Upgrade mechanism [139]. The client communicates with the server via the same TCP connection for the life cycle of WebSocket connection. (4) *Lean Protocol*: As the HTTP is a chatty protocol, a set of HTTP headers are sent in a request message by Advanced REST Client extension. However, the purpose of WebSocket is to break the limitations of the request/response protocol such as HTTP [139]. In summary, WebSocket provides an alternative approach to the REST/ SOAP/AJAX technologies for developing real-time communication web applications, like web-based control applications [138, 140]. Moreover, WebSocket is the next-generation method of asynchronous communication between the client and the server [141] and is already standardized by the World Wide Web Consortium (W3C). Currently, it is already implemented in the most of the popular web browsers such as Microsoft Edge, Chrome, Safari, Firefox, and Opera [139].

Socket.IO is designed based on WebSocket, and it enhances WebSocket by providing built-in multiplexing, horizontal scalability, automatic JSON encoding/decoding, and more [142]. In general, Socket.IO uses feature detection to decide which approach, such as WebSocket, AJAX long polling, or Flash, will be used to establish the connection for real-time web applications. Through blurring the differences between the different transmission mechanisms, it is possible for Socket.IO to support real-time web applications in any popular browsers. Socket.IO includes two parts: a client-side library for the browsers and a server-side library supported by Node.js. Although Socket.IO works as simply a wrapper for WebSocket, it provides more features including broadcasting to multiple sockets, storing data associated with each client, and asynchronous I/O, to support the development of real-time web applications.

To compare the performances of different real-time data transmission protocols, we tested 10, 100, 250, and 500 data exchanges per millisecond between the client web module and the server-side middleware in our new system. Each data exchange between the Node. js server and the Chrome browser is a 4K bytes random data string. The Node.js server is running in release model, without debug mode. Meanwhile, the output console messages are minimized for both the server and client. The server is HP Proliant DL380e Gen8. Hardware of the server includes Intel Xeon E5 2.5 GHz processer and 16GB of RAM. The network is the University of Houston's main campus Wi-Fi network. The download speed is around 45 Mbps and upload speed around 75 Mbps. The Socket.IO is a better data transmission module than WebSocket for real-time data communication without extra plug-ins. Consequently, Socket.IO is the best selection for us to be used for implementing real-time data transmission between the users with the experimental equipment.

2.3 THE DESIGN OF THE NOVEL FLEXIBLE FRAMEWORK

The novel flexible framework is designed based on the online IM application architecture, and a new-version LtoN module is developed to rapidly and flexibly integrate the RL. Figure 2.1 shows the architecture of the new flexible framework. The new flexible framework includes three parts: client web module, server module, and LtoN module for experimental control.

2.3.1 Client Web Module

In the design and implementation of the client web module, the MVC (model–view–controller) software architectural pattern is used for implementing the user interfaces (UIs) as shown in Figure 2.1. As client web technology has matured, frameworks such as AJAX, JavaScript MVC, and AngularJS have been created which allow the MVC components to execute partly on the web browsers. To run the client web module on any popular browser, the HTML5 technology is used for the web application implementation. Some popular development languages, such as HyperText Markup Language (HTML), Cascading Style Sheets (CSS), and JQuery/JQuery-Mobile JavaScript libraries, for web application development are involved as well.

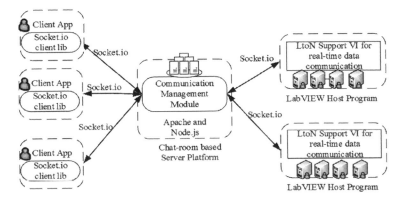

Figure 2.1 Novel flexible framework architecture.

In addition, to resolve the need of installing extra plug-ins for supporting efficient real-time data transmission, the Socket.IO module is used in client web module implementation. Meanwhile, the server-based mash-up technology is used for UI integration as well.

2.3.2 Server Module

To rapidly and flexibly connect the users with experiment equipment together, a communication management module is designed based on the online IM application architecture. Every user, who is a real person or an experimental equipment, is assigned a unique ID, and the system creates a communication thread and pairs them together based on this unique ID. With this module, the RL system based on the new flexible framework can flexibly pair one user to one-equipment experiment, multiple users to one-equipment cooperative experiment, one user to multi-equipment experiment, or multiple users to multi-equipment cooperative experiment. It will significantly benefit the RL implementation.

To resolve the web service technology performance issue, a combined solution of both Apache web engine and Node.js web engine is implemented for real-time communication between experimental equipment and end users. Node.js enables web developers to create a real-time-communication web application in JavaScript which is both server-side and client-side. In the Node.js server-side software system, Socket.IO, a JavaScript library, is used to support real-time communication between the server and the client. Moreover, a real-time video transmission solution based on the HTTP Live Streaming (HLS) protocol is also implemented in server side. Thus, the server module is directly built on top of an Apache web server engine, a Node.js web server engine, and a MySQL database. In addition, the operating system (OS) of the server uses Centos 7.0 OS to better support the server module.

2.3.3 The New-Version LtoN Protocol

To rapidly connect the experimental equipment into the RL platform, an essential issue of "How to design a real-time communication module which can be easily used to connect experimental equipment and end users?" must be addressed. To address this issue, an easy-to-use, real-time communication module based on Socket.IO, namely LtoN, is proposed and implemented using LabVIEW [3–4]. To fulfill the requirement of the new framework based on online IM application architecture, the unique ID management function has been integrated into the LtoN module. The new-version LtoN module can support P2P and M2M real-time communication via the unique ID assigned by the server. Moreover, some bugs have also been fixed in this new-version LtoN module to improve its performance.

NI LabVIEW software is one of the popularly deployed technologies for remote panel over the Internet. With the development of computer technology, LabVIEW also integrates a new feature to interact with the experiment Virtual Instruments by using RESTful web services technology. REST (Representational State Transfer) provides a lightweight protocol accessible to a wide variety of clients. The architecture does not require complex message transfer and provides a simple interface for a user to begin using web services in LabVIEW. However, it requires the client interface to be developed using different technologies, and LabVIEW plug-ins must be installed in the browsers. To resolve this plug-in issue and support efficient data transmission, the new real-time communication module LtoN is designed and implemented.

To design the LtoN module, the MVC design approach is also used. As shown in Figure 2.1, the LtoN includes a real-time data transmission controller and an experimental data transfer model. With the new-version LtoN, the experimental equipment control module, which implements the experiment control logic, is connected with Socket.IO module. In order to achieve the secure data transmission, a new message packet approach is designed and implemented. With this new-version LtoN module, the client web module can communicate real-time with experimental equipment without LabVIEW plug-ins. Meanwhile, any experimental equipment controlled by LabVIEW can be rapidly integrated into the RL system anywhere.

2.4 SUMMARY

In this chapter, a novel flexible framework has been designed and implemented successfully with a combination of advantages of the social online IM application architecture and the new-version LtoN module for experimental equipment real-time control. With the real-time communication solution free of software plug-ins and based on the innovative LtoN module, it effectively reduces the efforts to shorten the time of RL system integration. By assigning a unique ID for the existing experiment, it provides a more flexible approach to integrate the existing experiment into the RL system. With the new framework, a more powerful online distributed RL system has been developed. It offers a more flexible way to build up different pattern remote experiments for engineering education, research, and industrial applications.

Although the new flexible framework delivers a new rapid integration approach to support RL development, further development is still required to improve the stability and usability of the new framework. More specifically, areas that need improvement are as follows:

- Integrating the new flexible framework into more LMSs. Currently, we have implemented our Wiki-based RL management platform based on this new framework, and it includes a scheduler, a communication management module, a user management module, and a module for learning materials management. In future, we plan to integrate the new flexible framework into some popular LMSs (e.g., Moodle, MIT iLab shared architecture) to avoid double registration of students.
- Integrating some industrial experiments based on our new flexible framework, such as industrial equipment remote monitoring application and industrial equipment remote training application.

Compared with the traditional integration approaches of RL systems, the new flexible framework can also provide more innovative integration pattern for remote experiments and give users a more flexible selection to build up their own specific RL without the limitation of space.

3

CLOUD-BASED RLaaS-FRAME FRAMEWORK FOR RAPID DEPLOYMENT OF REMOTE LABORATORY SYSTEMS

3.1 INTRODUCTION

As discussed in Chapters 1 and 2, a generalized remote laboratory (RL) system is based on the browser–server architecture [5, 6] with three modules normally: a client web application in users' devices, a middleware which is usually a Learning Management System (LMS) on the server, and an experimental equipment control application in the workstation. However, a common denominator of most existing RL systems, whether for academic or for industrial purposes, is that they offer stand-alone solutions with limited or no capability to cooperate with other platforms [19, 20]. To easily share RL experiments on large-scale incompatible platforms, a standardized cloud-based framework is crucial. There are several critical issues that have been pointed out by research works in [143–145], and they can be listed as follows:

- How to achieve elasticity and the appearance of infinite capacity available on demand with automatic allocation and management?
- A unified interface to provide integrated access to cloud computing services is nonexistent.
- How to create third-party cloud services based on high-performance content delivery over commercial cloud storage service?

The revolutionary progress of cloud computing technology provides an effective solution for these essential issues. Recently, cloud computing has become increasingly popular due to its unique advantages over traditional computing models [146]. The National Institute of Standards and Technology (NIST) makes a comprehensive description in identifying cloud computing technology with parameters [147]. Based on the definition of cloud computing model from NIST, the Software as a Service (SaaS) is the first layer of standard cloud. The Platform as a Service (PaaS) is the second layer of standard cloud, and the Infrastructure as a Service (IaaS) is the third layer of standard cloud [147]. These three layers provide three different series of services to various users. Cloud computing adopts concepts from service-oriented architecture (SOA) and provides a new approach to supporting the system developers to break all of the functional modules of the system into services [148]. Cloud computing model incorporates the well-established standards and best practices, which are gained in the domain of SOA, to allow global and easy access to cloud services in a standardized way [149, 150]. Web service, which has a well-defined set of implementation approaches, is the most typical example of an SOA. However, the real-time data communication is still an essential challenging for cloud-based systems, and it has been an active area of cloud computing research [151, 152]. **Therefore, how to design a standardized cloud-based framework to rapidly and flexibly deploy an RL system still has not been resolved very well** [8–11]. Building of RL systems on a unified and standardized RLaaS-Frame (RLaaS: Remote Laboratory as a Service) can be easily deployed, and all resources can be shared and accessed by users with great convenience. Consequently, the RLaaS-Frame will significantly simplify works needed to integrate and implement RL systems on a large scale. To demonstrate the feasibility of the RLaaS-Frame, a wiki-based RL platform, and the mobile-optimized application framework has been integrated into the RLaaS-Frame. Meanwhile, this new RLaaS-Frame can be applied for the integration of industrial equipment remote control and monitoring applications as well. It will be a significant improvement for the development of RL systems technology in the future.

3.2 RLaaS-FRAME ARCHITECTURE

To address the vital issue, "How to design a standardized cloud-based framework to rapidly and flexibly deploy RL system?" The RLaaS-Frame is designed as a standard framework based on cloud computing model. The RLaaS-Frame originates from promoting the traditional

RL technology to the new cloud-computing-based RL technology. The RLaaS-Frame architecture is divided into three part services: experimental application (experimental applications) part services, platform (experimental development and integration platform) part services, and resources (experimental data, experimental video, etc.) part services.

3.2.1 Experimental Application as a Service (EAaaS) Layer

The Experimental Application as a Service (EAaaS) layer is the first layer of the RLaaS-Frame. The users of this layer use RL applications for education and research. To meet the needs of different users, the RLaaS-Frame offers different user interfaces (UIs) through this layer. To simplify the user's operations and reduce the efforts and cost of system maintenance, we adopt the browser–server architecture applications to provide the EAaaS layer service to end users. It supports most of the popular web browsers such as Chrome IE, Safari, and Mercury, which are running in different terminal devices (desktops, laptops, mobile phones, tablets, etc.). The goal of our design for the EAaaS layer is to provide flexible services to different end users through different RL applications. Experimental applications mainly include the following:

- RL applications (RLaaS-Frame-based application) for education
- RL applications (RLaaS-Frame-based application) for research
- Other applications (RLaaS-Frame-based applications).

3.2.2 Experimental Development Framework and Running Environment as a Service (EFEaaS) Layer

The Experimental Development Framework and Running Environment as a Service (EFEaaS) layer is the second layer of the RLaaS-Frame. The users of this layer design and develop their customized RL framework and architecture by using virtualized technology. They will gain RL development and integration support services from this layer. From the RLaaS-Frame architecture, the RLaaS-Frame provides a remote experiment development and integration framework and platform-level running environment (such as the series of application programming interfaces (APIs), the interface of database, and the interface of experimental data analysis component and tool kits). Users can develop their customized RL solutions based on the APIs that are provided by the EFEaaS layer without too much effort expended in developing and maintaining the complex system software layers and managing the underlying hardware. With services of the EFEaaS layers, the underlying hardware and storage resources automatically match the requirements of applications internally, and users don't need to be concerned with the procedures of resource management and maintenance. Experiment integration/development framework and running environment mainly include

- Experimental application framework for secondary development
- Experimental data transmission protocol
- Experimental data analysis method and algorithm component
- Experimental data processing toolkits.

3.2.3 Basic Experimental Resources as a Service (BERaaS) Layer

The Basic Experimental Resources as a Service (BERaaS) layer is the third layer of the RLaaS-Frame. The users of this layer are those who request experimental equipment setup and configuration services, basic experimental resources, and environment delivery services for their customized RL system. From the RLaaS-Frame architecture, the BERaaS layer offers the basic experimental units (OS, VM, software for experimental data collection, configuration of experiment server computers, configuration of network cameras, configuration of experimental equipment, etc.), experimental equipment control software components, and resources (experimental data, experimental videos, etc.). Basic experimental resources mainly include the following items:

- Experimental unit cluster management
- Experimental unit duplication sharing
- Experimental environment setup and configuration image duplication
- Experimental equipment control component
- Collecting and storage experimental data
- Experimental video.

3.2.4 Characteristic of the RLaaS-Frame

As RLaaS-Frame is based on the cloud computing technology, it not only comprises certain standard characteristics from the cloud computing technology, but it also has some of its unique special characteristics. The specific characteristics of the RLaaS-Frame are as follows:

- The accessibility of standard APIs of the RLaaS-Frame to software enables machines to interact with RLaaS-Frame-based platform in the same way as the UI facilitating interaction between users and computers. The RLaaS-Frame consists of a set of standard APIs with REST (Representational State Transfer) based on SOA WebService architecture. The standard APIs are mainly utilized to support the cloud-based applications' development and integration.
- Device and location independence enable users to access systems only using a web browser regardless of their locations or devices they are using (desktops, laptops, tablets, pads, mobile phones, etc.). As infrastructure is off-site (provided by different physical laboratories in different places worldwide) and accessed via the Internet, users can connect with the RLaaS-Frame-based RL platform from anywhere and at any time.
- Virtualization technology allows for easy and smooth sharing and utilization of servers and storage devices which are in physical laboratories in different locations. It is easy to migrate RLaaS-Frame-based applications from one physical server to another (even though they are in different locations worldwide).
- Security performance in RLaaS-Frame-based remote laboratory platform is often better than or at least as good as other traditional RL systems, in part because the RLaaS-Frame-based RL service provider can centralize management of experimental data and increase security-focused experimental resources. Meanwhile, the service provider can also establish the unified security passcode in the standard experimental data transmission protocol which is developed in the RLaaS-Frame to control experimental data security. However, the complexity of security is greatly increased when experimental data is distributed over a wider area or a greater number of devices. Consequently, the security management module of the

RLaaS-Frame needs to be constantly upgraded based on the solutions to different specific security issues.

• Maintenance and updates of the RLaaS-based platform become easier. As the end users will only use web browsers without any software plug-in to access the RLaaS-Frame-based RL platform from anywhere in the world, they do not need to worry about software update issues. The RLaaS-Frame-based platform only needs to perform maintenance on the server side.

3.3 THE REFERENCE DEPLOYMENT PROCESS OF THE RLaaS-FRAME

Based on the definition of the RLaaS-Frame, an SOA has been built based upon the cloud computing technology. Real laboratory resources (experimental equipment, experimental data, experimental control software, experimental methods, etc.) are delivered with the standard protocol as a service over a network (typically the Internet). The end users use the RL services without the need to understand how the system and infrastructure work with the new RLaaS-Frame-based RL system.

With the cloud computing technology, the new RLaaS-Frame architecture is implemented successfully to provide the RL services for end users. In Figure 3.1, we can see the features of the new RLaaS-Frame-based RL system. The whole RLaaS-Frame includes three parts: resource aggregation service, platform and framework service, and application service.

3.3.1 Application Service

Based on the EAaaS layer of the RLaaS-Frame, some different remote experimental applications provide different RL services for academic, industrial, and research activities. Most of the applications are deployed in Docker containers managed by Kubernetes. Kubernetes can facilitate automated container deployment, scaling, and management. The applications consist of three components: web server (Apache web server, Nginx web server, etc.), database (MySQL database, MongoDB, etc.), and experiment service package (Node.js, HTML5, RESTful API, etc.). The applications can run on most of the popular browsers and mobile platforms, and are built on experiment service package. A cluster of applications can be created with the simple drag-and-drop interface with OpenStack Horizon. OpenStack

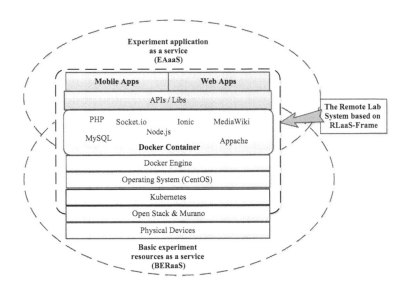

Figure 3.1 Deployment of RL systems.

orchestration Heat creates all the resources, networking, and computing needed for the applications. All applications are containerized as Kubernetes Pods which are generated in the Kubernetes cluster environment.

3.3.2 Platform and Framework Service

The RLaaS-Frame provides platform and framework services through OpenStack, Murano, Kubernetes, and Docker technology. To integrate Docker into OpenStack cloud environment, two keystone technologies are involved: the Murano and Kubernetes. Murano is adopted to compose reliable application environment and publish various cloud-ready applications. OpenStack works as the underlying layer and provides basic cloud computing services. Kubernetes is introduced to manage container clusters, discover services, and isolate resources. Platforms and frameworks in the RLaaS-Frame are deployed in Kubernetes clusters based on OpenStack using Murano. Figure 3.1 shows all of the nodes in the platform and framework service, which include the unified framework, the MediaWiki platform, and mobile-optimized architecture. An adapter API layer for data exchange is required for data communication between the BERaaS layer and the EFEaaS layer. To address the real-time data communication challenge of cloud-based systems, Socket.IO protocol is used to implement real-time experimental data transmission between the EFEaaS layer and the EAaaS layer. Socket.IO is designed based on WebSocket and enhances the WebSocket by providing built-in multiplexing, horizontal scalability, automatic JSON encoding/decoding, and more. All the components including Nova, Heat, Keystone, and Horizon in OpenStack work together with Murano to deploy the Kubernetes. Each Kubernetes node implements different platforms and frameworks with multiple Kubernetes pods. Kubernetes pods are Docker containers installed with software packages and components of RLaaS-Frame services. Murano provides the application catalogue, cluster management, and infrastructure automation options for the EFEaaS layer to facilitate the fast deployment of platform and framework services. Considering the security cloud environment, Kubernetes also provides the portability to be isolated from the BERaaS layer.

3.3.3 Resource Aggregation Service

There are two key points in implementing the resource aggregation service: data acquisition and video stream. The STEM (science, technology, engineering, and math) education consists of sensor-driven experiments that can deploy the RL based on the BERaaS. The experimental device needs to communicate with EFEaaS layer via the unified framework APIs. Data transmission and video streaming are the basic services that are needed to transfer and present data to end users. The resource aggregation service provides the standard protocol for data acquisition, data transmission, and video streaming. To enable real-time video streaming, the HLS (HTTP Live Streaming) protocol is adopted in the resource aggregation service. The Socket.IO protocol can deliver real-time data transmission combined with the Node.js software package. Resource aggregation service is the intermediate between physical laboratory and cloud computing.

3.3.4 Detail Reference Deployment Process

To prepare the system hardware and network for the implementation of an RL system, the HP ProLiant DL380e Gen8 server and experimental network have to be set up. The hardware of the server includes Intel Xeon E5 2.5 GHz processor and 16 GB of RAM. Other pieces of equipment include Axis web cameras, HP workstations, and a Cisco router. To build up an experimental network, the web cameras and workstations have to be connected to PoE (Power over Ethernet) of router switch through four Ethernet cables, and one Ethernet cable from the PoE switch connects to the server.

With the RLaaS-Frame, the RL system will provide experimental services to end users. As shown in Figure 3.1, the first step is to review the RLaaS-Frame model and then prepare the software package for the system's deployment. As shown in the figure, the system package includes (1) a new assembled server engine scheme, which consists of two server engines, Apache HTTP server engine and Node.js server engine; (2) real-time communication modules, which include a real-time experimental video transmission module and a real-time communication module; (3) MediaWiki engine and Ionic framework used to support the different remote experiment apps, which include web app and mobile app; and (4) MySQL used for database management system. To install and configure the system package, all of these components are packaged in different Docker containers. Kubernetes cluster was implemented including the Docker containers. Combined with OpenStack Murano, the experiment's essential services like MySQL and Apache can be integrated into OpenStack via the Kubernetes orchestration engine. Based on OpenStack, the RLaaS-Frame can implement the Wiki platform and mobile-optimized remote experiment architecture with better cloud computing resources. The RLaaS-Frame can embrace the flexibility of OpenStack by using more powerful tools like compute, storage, and networking. After the server system deployment, we have to configure computer node manager and server migration. In the end, the RL services will be offered to end users. The RLaaS-Frame promotes an environment for the more agile deployment of RL system.

To integrate the RL platform and framework into cloud environments, as shown in Figure 3.1, there are three steps: deploy OpenStack service, create Kubernetes cluster using Murano, and implement the platform and framework in Kubernetes. OpenStack contains functional APIs to integrate with Murano. Murano serves as the cluster conductor node to deploy the Kubernetes cluster by using the Horizon UI. Docker orchestration engine can manage Docker containers in Kubernetes cluster. Docker containers are installed with software packages (Apache web server, MySQL, Node.js) to implement the applications.

3.4 SUMMARY

In this chapter, a cloud-based framework, namely RLaaS-Frame, was successfully established for rapid deployment of RL systems. With this new RLaaS-Frame, the real-time data communication challenge discovered in the RL system implementation can be solved as well. As a pilot RL platform based on the RLaaS-Frame, a Wiki-based RL platform and the mobile-optimized application architecture have been integrated into the RLaaS-Frame successfully.

However, further research and development are still required to improve the RLaaS-Frame and enhance the stability and usability of the RLaaS-Frame. More specially, issues that need improvement are as follows:

- Various experiments for different disciplines need to be designed following the standardized RLaaS-Frame. With a growing number of new requirements from users, more and more new remote experiments need to be designed and implemented based on the RLaaS-Frame in future.
- Artificial intelligence technology (AIT) and big data technology will significantly improve RL development in such a way as to enhance mobile learning and online learning. The use of the new RLaaS-Frame provides more effective RL services to users; it can be used to support the cloud-based remote experiment for various purposes in different fields.

With the new RLaaS-Frame, RL systems have been quickly deployed successfully. Users can apply this cloud-based RL platform to learn engineering knowledge anywhere and anytime without plug-in issues. The RLaaS-Frame will significantly benefit online engineering education, academic research, and industrial applications.

4

A NOVEL MOBILE-OPTIMIZED REMOTE LABORATORY APPLICATION ARCHITECTURE

4.1 INTRODUCTION

As discussed in Chapters 1 and 2, to offer users a more flexible way to access RL systems, instead of forcing the users to sit in front of a fixed computer to use a location-independent environment for experimentation, a technology that is suitable for presenting remote laboratory (RL) applications on mobile devices becomes essential. In addition, integrating an RL into mobile devices can offer users a more flexible approach to learning and produce better outcomes as pointed out by May et al. [87] and Silva et al. [88]. So far, most of current research interests in mobile learning have mainly focused on the various learning theories; only a smaller number of researches have focused on the design of framework and the mobile device technologies that are compatible for M-learning systems [89, 90]. **How to design and implement a mobile-optimized and easy-to-use application for M-learning has become an emerging research topic** [91, 92]. Generally, two approaches, the native mobile application and the web-based application, are used to integrate an RL into mobile devices.

Therefore, **how to design a mobile RL application architecture, which has native-like performance and native functional capability, can also run cross-platform easily, and needs low cost and less development efforts like web applications, is already an essential issue** [12–14]. To address this essential issue to better support M-learning, a new mobile-optimized RL application architecture based on Ionic framework is proposed and implemented in this chapter. To the best knowledge of the authors, this is the first study to build a mobile-optimized application architecture for RL application development based on Ionic framework.

This new mobile-optimized RL application architecture can provide an innovative development tool for mobile industrial electronic devices with less effort as well.

4.2 NOVEL MOBILE-OPTIMIZED APPLICATION ARCHITECTURE

To answer the research question, "How to develop a mobile RL application running cross-platform easily with the competitive running performance and hardware accessibility of a native application?" a new mobile-optimized application architecture is proposed. This new architecture which combines Ionic framework with the unified framework together can have the advantages of both frameworks. The mobile-optimized application architecture includes two layers, optimized application layer and unified framework layer, as shown in Figure 4.1.

4.2.1 Mobile-Optimized Application Layer

The mobile-optimized application layer can support the running of applications on most of popular mobile platforms. As this layer is implemented based on Ionic framework, it also works on two

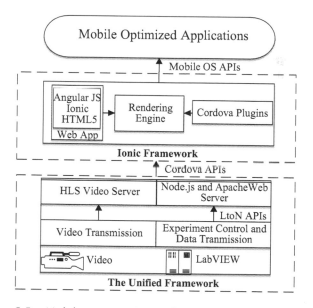

Figure 4.1 Mobile-optimized RL architecture.

kernel components, Apache Cordova and AngularJS. (a) *Apache Cordova*: Apache Cordova is an open-source mobile development framework. To avoid each mobile platform's native development language, it allows the developer to use standard web technologies (such as HTML5, Cascading Style Sheets (CSS3), and JavaScript) for cross-platform development. Mobile applications are executed within wrappers on different platforms, and they rely on the standard application programming interfaces (APIs) to access the different devices' sensors, data, and network status. To maximize the native mobile devices' hardware capabilities, the Apache Cordova framework, as a plug-in, is integrated into the new application architecture. (b) *AngularJS:* AngularJS is a structural framework for dynamic web applications' development. Developers can use HTML as the template language and extend HTML's syntax to express an application's components clearly and succinctly. It also provides the data binding function and dependency injection function to eliminate the redundant code of the target mobile application. As more and more new APIs are constantly developed for the interactive features of mobile applications, the Ionic framework supports more and more mobile systems, e.g., Android, iOS, Windows Mobile, Blackberry, Amazon Fire OS, Firefox OS, Ubuntu Mobile OS, and Tizen [153]. To enhance the mobile-optimized application performance, Crosswalk as the rendering runtime engine is integrated into the mobile-optimized application. Moreover, Crosswalk also runs as a runtime engine in different mobile systems to automatically update the rendering engine based on different platforms. To develop an easy-to-use application user interface (UI), the optimized application layer relies on standard APIs of WebView for the application presentation layer. As the WebView can effectively improve the performance and user experience, it works as a middleware between the web technology (such as AngularJS, HTML5, and Apache Cordava) and native mobile systems. Depending on the different frameworks (such as AngularJS and Apache Cordava), many different types of WebViews can be used for mobile application implementation. The detail modules of the optimized application layer are listed as follows:

- Apache Cordova module
- Web Application support module (implemented with AngularJS, Ionic framework, and Common Codebase)

- WebView (cross-walk rendering runtime engine)
- Mobile systems' native APIs module (Android, iOS, Windows Mobile, etc.).

4.2.2 Unified Framework Layer

In order to integrate the RL technology into the new mobile-optimized application architecture, the unified framework is merged into this new mobile application framework. The unified framework is based on the combination solution of both Apache web engine and Node. js web engine, and it uses a Node-HTTP-based technology to implement real-time communication between experimental equipment and end users. The subsequent iteration of the design resolved challenges of developing cross-browser and cross-device web UI as an improvement to the unified framework. To implement real-time communication between experimental equipment and mobile application without plug-ins, a real-time communication module based on Laboratory Virtual Instrument Engineering Workbench (LabVIEW) to Node.js (LtoN) protocol and an experimental control program are developed using LabVIEW which stands for Laboratory Virtual Instrument Engineering Workbench.

National Instruments (NI) LabVIEW software is one of the popularly deployed technologies for remote panel through the Internet [63]. With the development of computer technology, LabVIEW also integrates a new feature to interact with the experiment Virtual Instruments by using RESTful web services technology. REST (Representational State Transfer) provides a lightweight protocol accessible to a wide variety of clients. The architecture does not require complex message passing and provides a simple interface for a user to begin using web services in LabVIEW. However, it requires the client interface to be developed using different technologies, and LabVIEW plug-ins must be installed in the web browsers [154]. To resolve this plug-in issue and support the new mobile-optimized RL application architecture, the new application communication module was implemented based on the LtoN protocol. This module includes three parts: a client part that runs in mobile-optimized applications, a server part that runs in the Node.js web server, and a control module that runs in LabVIEW experimental equipment control program. The client part and server part were developed with JavaScript language, and the control module was developed in LabVIEW.

With this new application communication module, the mobile-optimized application can communicate real-time with experimental equipment without LabVIEW plug-ins in the client side. The detail modules of unified framework layer are as follows:

- Real-time video transmission module (implemented based on HTTP Live Streaming (HLS) protocol)
- Real-time experimental data transmission module (implemented based on LtoN protocol based on web socket)
- Combinational Server Engine (Node.js web server and Apache web server)
- Common Codebase (HTML, JavaScript, CSS)
- Data Management System (MySQL)
- Experimental equipment control module (implemented by LabVIEW Virtual Instruments (VIs)).

4.2.3 Characteristics of the New Mobile-Optimized Application Architecture

In Figure 4.1, we can see the detail characteristics of the new mobile-optimized application architecture. As the Apache Cordova framework supports cross-platform application development, the mobile-optimized architecture can support most of the popular mobile systems. In addition, since the AngularJS framework can support development of dynamic web applications, lots of native features of different mobile systems can be realized based on the new mobile-optimized application architecture. WebView is also the key technology in this new architecture to support reusing the common codebase. Therefore, developers can deploy the mobile-optimized applications without spending much time on rewriting the code for different mobile platforms based on the new application architecture.

4.3 SUMMARY

In this chapter, a new mobile-optimized application architecture is designed and implemented successfully to provide a new tool for development of applications in many fields, such as M-learning application and industrial electronics applications. It integrates the advantages of both native mobile applications and web applications. Meanwhile, it improves the running performance and solves hardware accessibility issues of web applications. The issue of native applications cross-platform running is solved as well. Moreover, it seamlessly combines the unified framework and Ionic framework together to deliver excellent RL services to end users.

Although the new mobile-optimized RL application architecture delivers a new development tool to support student-centered mobile learning, there are still further developments required to improve the new architecture's stability and usability. More specifically, issues that need improvement are listed as follows:

- Integrating the new mobile optimized RL application architecture into Learning Management Systems (LMSs). Currently, we have developed our own RL management platform, and it includes a scheduler, a user management module, and a module for learning materials management. In the future, we plan to integrate our mobile-optimized RL application architecture into an open-source LMS (e.g., Moodle) to avoid double registration of users.
- The new mobile-optimized RL application architecture is still version 1.0. So far, only a few sample pilot tests have been conducted. Some bugs in the software package have been fixed through user feedback. The future work will be to further refine and improve this mobile-optimized application architecture for M-learning.

5

A NOVEL ONLINE PROGRAMMABLE PLATFORM FOR REMOTE PROGRAMMABLE CONTROL EXPERIMENT DEVELOPMENT

5.1 INTRODUCTION

As discussed in Chapters 1 and 2, traditionally, remote control experiments are designed to be used on a specific topic: controller tuning, stability, verifying a control model, etc. Users can utilize the basic control rules only through changing a set of modifiable parameters which are built in the controller to get the corresponding result. Although these traditional remote control experiments have been working satisfactorily for over decades for gaining basic control knowledge, they are still very difficult for users to develop and verify their own designed control algorithms and models remotely [15, 16]. Nowadays, Easy Java/JavaScript Simulations (EjsS) provides a chance to overcome this problem [16, 155]. Users can write the customized code to implement their own designed control algorithm and model through the web application supported by EjsS server engine with java applet plug-in. However, the EjsS only supports the server-side programmable environment for Virtual Labs [16].

Consequently, **how to develop an online programmable platform to support users to implement their own designed control algorithm and model to control the real experimental devices remotely real-time** has been an emerging topic [15–18]. In this chapter, a novel online programmable platform is proposed, and its successful development is discussed. This new online programmable platform utilizes the advantages of the Wiki-based remote laboratory (RL) platform [26] and web-based integrated development environment (IDE) for engineering hands-on practice. To the best of my

44

knowledge, this is the first online programmable platform for remote programmable control experiment development to be presented with a collaborative learning environment. With this new online programmable platform, users can collaboratively design and implement new remote programmable control experiments to engage users in engineering education more effectively.

To discover a suitable solution for the proposed online programmable platform, I have to understand which part of the traditional RL system offers programmable capability, and select a suitable online IDE for the online programmable platform. Based on the remote experiment's general architecture [6], three components can provide the programmable capability to offer users programmable hands-on experience: the programmable experimental device, the software in workstation, and the server-side programmable environment to control experimental devices.

5.1.1 Programmable Logic Controllers for Experimental Device Control

Programmable logic controllers (PLCs) are computer-based, solid-state, and single-processor devices that emulate the behavior of an electrical ladder diagram and are capable of controlling many types of entirely automated equipment [156]. PLC control systems have been designed to be easily installed and maintained. Trouble shooting is simplified by the use of fault indicators and messages displayed on the programmer development environment. Input/output (I/O) modules, which connect to the field devices, are easily connected and replaced [157]. Generally, PLCs are designed to be programmed with schematic or ladder diagrams instead of common computer languages [158–160] and are widely used in automation control in industry. The common working process of programming a PLC is to design the desired control circuit in the form of a relay logic ladder diagram and then to load this ladder diagram into a programming terminal. The programming terminal is capable of converting the ladder diagram into digital codes. Then the developed program is sent to the PLC where it is stored in memory.

To implement the remote control experiments, a control unit, PLC, is one of the popular approaches. Normally, the control algorithm model code is directly loaded in the PLC memory. Almost all of the common features, like data acquisition, DC motor variable control, and connecting with control PC, are handled by PLC hardware. Recently, there are many successful examples implemented using

PLCs, such as a remote SCADA control experiment with PLCs [161], a motor control experiment with PLCs [162], and a distributed peer-to-peer (P2P) remote control experiment with PLCs [117]. However, the drawback of PLC programming is the complex development process, and it is difficult for users to learn and master. Thus, all of current remote control experiments based on PLC only offer users hands-on experience in changing the characteristics and parameters of visualizing the process of remote control experiment.

5.1.2 Software in Workstation for Experimental Device Control

There are many software tools for experimental device control, such as LabVIEW [5], MATLAB [163], Simulink [164], and Scicos [165]. In these tools, Laboratory Virtual Instrument Engineering Workbench of National Instruments (NI LabVIEW) is almost the most common software system for experimental device control application development [6], [26]. NI LabVIEW is one of the popularly deployed technologies for a remote panel over the Internet [63]. LabVIEW integrates several new features to interact with the experimental Virtual Instruments (VIs) by using RESTful web services technology. Representational State Transfer (REST) provides a lightweight protocol which is accessible to a wide variety of clients. The architecture does not require complex message passing, but rather it provides a simple interface for a user to begin using web services in LabVIEW. However, it requires the development of the user interface (UI) using different technologies, and LabVIEW plug-ins must be installed in the web browsers [154].

The control algorithm and model is normally developed using software tools in PC workstation. Currently, because most of RL systems use the LabVIEW as the control software tool, all the functions of the designed control algorithm and model could be implemented by LabVIEW VIs which have been executed in the workstation. Then, the control algorithm and model can communicate real-time with the experimental devices through the workstation. This approach requires users to be familiar with the control software tool at first, and then they can implement their own control algorithm and model with the software tool. Although this approach is easier than PLC programming in the development process, the users must implement the code in the workstation. It is still hard to support programming remotely and flexibly offer hands-on experience for users. Consequently, most of the remote control experiments based on control software tool only offer students hands-on experience to develop their own designed control algorithm and model by changing the characteristics of visualizing the process of remote control experiments.

5.1.3 Server-Based Programmable Environment for Experimental Equipment Control

The main limitation of traditional RL architecture is that users can only manipulate certain parameters of the experimental equipment without accessing the core of the controller to change the control model. To offer a remote access platform, which allows users to implement their own control algorithms and models, such as PID and sliding-mode controller. Temelta et al. have proposed a solution to control an open-architecture 6-DOF PUMA 560 manipulator and two hardware-in-the-loop (HIL) robotic simulator platforms at the UAF Remote Robotics and Control Laboratory [17]. The users can use these test beds via web browsers on the client-side computers. The connection between the remote user and the experiment is provided by a database server [166]. Another platform allowing for users' own algorithms is the Networked Control System Laboratory in the University of Glamorgan, U.K. [18]. The platform has allowed users to implement their own control algorithms for the test rigs using MATLAB Real-Time Workshop and a template file. The web interface of this platform is developed by Java JSP/Servlet [18]. However, in these programmable platforms, the MATLAB/Simulink plug-ins have to be installed in client-side web browsers. With the continuous development of RL technology, EjsS provides a chance to support the server-side programmable environment for Virtual Labs [155]. Users can write the code via web browsers to implement and test their own designed control algorithm and model through the Java applet plug-in provided by EjsS server-side application.

5.2 ONLINE PROGRAMMABLE EXPERIMENT PLATFORM ARCHITECTURE

The online programmable platform aims to offer a new online programmable experimental environment for users to implement and verify their own designed control algorithm, which is running in server side, and the code is input via the web browsers without other software plug-ins remotely.

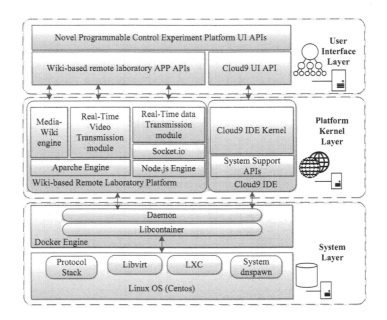

Figure 5.1 Proposed online programmable platform architecture.

Based on the related work, Cloud9, a web-based IDE, can provide the strong capability to move the program development progress from workstation to server side [167]. Meanwhile, with the Wiki-based RL platform, users can collaboratively develop new customized remote control experiments to support student-centered online engineering learning more effectively [26]. Thus, the essential issue to be addressed is, "How to develop an online programmable tool to support users to implement their own designed control algorithm and model to control the real experimental devices remotely real-time?" The integration of a Wiki-based RL platform and Cloud9 online IDE provides a suitable solution for the implementation of the proposed online programmable platform. The new online programmable platform architecture includes three layers: UI layer, platform kernel layer, and system layer as shown in Figure 5.1.

5.2.1 UI Layer

The UI layer is used to support the implementation of the client web applications. The new programmable control experiment web application runs on web browsers and is based on HTML5 and

Web 2.0 technology. Further, it uses the server-based mash-up technology for UI implementation. Cloud9 UI API and Wiki-based RL app APIs have been used to implement the new programmable control experiment app as well. Through the UI of the new programmable control experiment app, users can develop their designed algorithm and control model with their familiar program languages via web browsers without any plug-ins. Moreover, as the Wiki-based RL platform can provide strong social networking services, such as chat room, search engine, and blog, the new remote programmable control experiment app can provide strong communication and cooperative capability to create a better student-centered learning environment [26]. By adding the online programmable capability into the remote control experiments, the new platform can offer much stronger support for engineering education.

5.2.2 Platform Kernel Layer

In the platform kernel layer, there are four core modules: Cloud9 IDE kernel, MediaWiki engine, real-time data transmission module, and real-time video transmission module. To support MediaWiki engine and two real-time communication modules, a combined solution of both Apache web engine and Node.js web engine has been used [26]. Cloud9 IDE is the key module to build a server-side programmable environment, and it is supported by system-support APIs. Moreover, with the code parsing, building, and inline queries of Cloud9 IDE, the user-developed program running in server side can communicate with the experimental device in a real-time manner. To combine the MediaWiki engine and Cloud9 IDE kernel together, all of the modules in the platform kernel layer need to run in Docker Daemon. Docker has been deployed as the lightweight container to package the web-based IDE and other software modules [168]. As Docker container does not replicate the full operating system (OS), it is not the same as the traditional virtual machine, and it only virtualizes the libraries and binaries of the application [169]. Thus, Docker offers the ability to deploy scalable services, such as web-based IDE and MediaWiki engine, on a wide variety of platforms. Moreover, the container package with Cloud9 can be distributed into the Wiki-based platform without worrying about inconsistencies between development and complicated environments.

5.2.3 System Layer

The system layer has been used to support the platform kernel layer execution. In this layer, the Docker engine builds on top of the Linux OS, and Centos is used in the authors' system. Docker guarantees the applications run the same, regardless of the environment (computers, local servers, private and public clouds, etc.) where they are running in. Docker wraps up the software package in a complete file system that contains everything it needs to run. Docker Daemon runs on the Libcontainer, which is used to provide the execution environment. To successfully execute the components of the platform kernel layer, the execution environment must be built up and configured. Some important build-in execution drivers, such as Libcontainer, Libvirt, LXC, Systemd-nspawn, and protocol-stack, must be installed. As the Docker can image the full-fledged system layer; greatly reduces the burden of building, deploying, and maintaining the platform kernel and system layer; and only needs to update the UI layer based on users' various requirements.

5.2.4 Advantages of the Novel Online Programmable Experiment Platform

With the integration of the Wiki-based RL platform and Cloud9 online IDE into a full solution, this new online programmable platform can provide the advantages of these two software systems, Wiki-based RL platform and Cloud9 online IDE, to end users. Several advantages of using this new platform are listed as follows:

- Programmable capability: The new online programmable platform offers a flexible environment for users to implement and verify their own designed control algorithm and model.
- Community for collaborative learning: As the Wiki-based RL platform offers strong social networking services to build up a community, which includes lecturers, students, and researchers, the new programmable remote control experiment based on this new online platform can support collaborative learning and research very well.
- Rapid configuration for own designed experiment: Based on the new online platform, users can create the new remote programmable control experiment rapidly.

5.3 SUMMARY

In this chapter, a novel online programmable experiment platform has been proposed. To implement this new online platform, an open-source web-based IDE, Cloud9, has been integrated into the unified framework. This new remote programmable experiment platform offers an approach to freely control an experiment with a fixed control model and allows users to input their own designed control algorithm code via web-based UI without any other software plugins. In addition, it offers users an online tool to promote the active participation in lab practice as well.

However, there are still further tasks required to improve the stability and usability of this new online platform. More specifically, several issues that need to be improved are listed as follows:

- More specific programmable control experiments need to be developed. With more and more new requirements from users, increasingly new remote programmable control experiments need to be developed based on this new programmable platform in future.
- In 2010, James Kuner at Google introduced the term "cloud robotics" [170] to describe a new approach to robotics that takes advantage of the Internet as a resource for massively parallel computation and real-time sharing of vast data resources. Based on this new remote programmable robot control experiment, a new cloud robotics application development tool needs to be implemented to provide a more powerful tool for cloud robotics and robot control research.

Compared with a traditional remote control experiment, the new online programmable platform offers a more flexible way to support users' collaborative learning and research activities.

6

WIKI-BASED REMOTE LABORATORY PLATFORM FOR ENGINEERING EDUCATION

6.1 INTRODUCTION

As discussed in Chapters 1 and 2, the social computing technology has already been used in the online laboratory systems' implementation. Because the student is the center of any learning service, online course offerings must provide a unique value to maintain high-quality education instead of simply repackaging content designed for traditional face-to-face learning environments. To provide unique advantages for student-centered learning, Wiki technology was proposed to create a collaborative learning environment [171]. Wiki technology, which is one of the most common forms of social computing, allows multiple users to collaborate and cooperate in generating and customizing knowledge in a user-friendly editing environment. As the name implies, Wikipedia is a prime example of a collaborative environment allowing users to create an encyclopedia using Wiki technology. Supported by the Wiki engine and run by Wiki software, a Wiki allows users to edit a page or create a page of knowledge inside the Wiki. In most cases, most of the editing can be done directly through a web browser without additional plug-ins or strong programming skills. The ease of use along with the Wiki's intrinsic advantages as a combined knowledge manager and social platform is essential to creating a collaborative learning environment. Another benefit of a Wiki-based learning environment is that the Wiki can provide a much stronger social context that is lacking in the current state of online courses. The advantages of collaboration provided by the Wiki technology has shown to strengthen the building of social ties, development of constructive peer critiquing skills, and other essential social skills [172–174]. With Wiki technology serving as the stage for learning, students and instructors may easily collaborate and communicate. In the long term, they can also co-design the structure and direction of a course through a co-create knowledge process. Thus, compared to most current social environments (e.g., face to face and online), Wikis are able to add an extra collaborative and active learning dimension that greatly encourages a collectivist effort among students and instructors to construct new knowledge [175, 176]. As an important component of online learning, the remote laboratory (RL) technology also has made great progress. The concept of an RL implies the use of Internet and system control technologies to remotely conduct real-time experiments. This location-independent access is especially useful in scenarios where space is limited or for distance education [42]. As it was pointed out in [5, 6, 19], most of the improved RL solutions mainly focus on the technology innovation, such as software and hardware platform upgrade and use of new IT technology. **Developing a well-structured and coordinated online learning platform to improve engineering education becomes an essential issue** [20].

To address aforementioned issues, we propose a Wiki-based RL learning platform to utilize advantages of the Wiki technology and RL technology for online engineering education. Based on this novel learning environment, a collaborative and cooperative RL platform is created. To the best of our knowledge, this is the first RL to present the remote experiment with a collaborative and cooperative learning environment. Integration with Wiki technology is an essential improvement for RL development in future. Through the new learning environment supported by RL technology, students and instructors can collaboratively design and implement new experiments to support student-centered online engineering learning more effectively.

6.2 WIKI-BASED RL SYSTEM ARCHITECTURE

To seamlessly integrate the unified framework into the Wiki platform, a Wiki-based remote laboratory platform architecture (WRLPA) is proposed to implement a new Wiki-based RL platform. A schematic of the WRLPA is shown in Figure 6.1. To provide an excellent student-centered learning platform, three layers of software development will be implemented: (a) the database layer, (b) the platform layer, and (c) the client layer. At the bottom, the database layer includes a Data Pool (DP), which contains all the information such as learning materials, experiment data, student performance records, student blog, and communication records. This layer mainly provides data storage, data

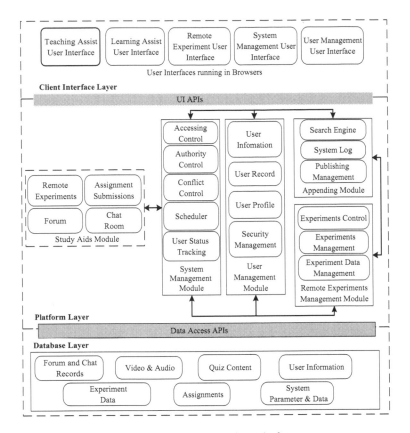

Figure 6.1 Architecture of Wiki-based RL platform.

retrieval, and query to support platform layer. The middle layer, the platform layer, which is the core of the Wiki-based RL platform, consists of the following modules: a study aids module (SAM), a remote experiments management module (REMM), a system management module (SMM), a user management module (UMM), and an appending module (AM). The SAM, REMM, and SMM work together to provide some new key functions of the new RL platform to support the student-centered collaborative and cooperative learning. The AM mainly provides a social context for students' learning. It means that students can communicate with others, search some learning materials, and search new remote experiments via this module. The UMM provides user management functions to support the normal usage of the systems. On the top is the client layer which includes different user interfaces to be supported by a set of user interface application programming interfaces (UIAPIs) that help the user interact with the platform. This layer provides a friendly user interface (UI) for students to support their different learning activities.

To achieve the goal of cross-platform deployment, the Wiki-based RL platform is implemented by Hypertext Preprocessor (PHP) language and JavaScript language. These two general-purpose scripting languages, which are especially suitable for server-side web development, will be run on a Linux, Apache, Node.js, MySQL and PHP (LANMP) web server. The detailed descriptions of these three layers are as follows.

6.2.1 Client Layer

In the client layer, there are mainly three different kinds of applications (i.e., teaching applications, learning applications, and system management applications) to support student-centered collaborative and cooperative learning for three different user groups (i.e., students, instructors, and system administrators). Users can create new contents (new remote experiments) easily by interfacing with the client layer. Different users may choose different styles of UI, and users are assigned different levels of authority to use the Wiki-based RL platform. Different users can only edit their own learning materials. As co-creating the remote-experiment-based learning or teaching materials is the core new function of this new learning platform, students can work with their fellow students to generate some creative remote experiments to support their learning. For the client layer design, the interface, logic, and content separation design pattern are used to ensure ease of maintenance and upgrading. An adapter API layer for data exchange is needed for communication between the platform and client layer. Socket.IO protocol will be used to implement the real-time experimental data transmission between the UI layer and the platform layer. Meanwhile, the Simple Object Access Protocol (SOAP) also is used to implement the transmission of other data (such as assignments data, and quiz data) between the UI layer and the platform layer. Data are saved in Extensible Markup Language (XML) format file for transmission.

6.2.2 Platform Layer

In the platform layer, there are two core modules, REMM and SAM. The SAM, REMM, and SMM work together to provide key new functions of the new platform to users, such as co-creation of a new remote experiment, generation of new experiment-based teaching or learning material, and creation of a cooperative remote experiment for a group of students. There are 5 modules and around 40 sub-modules that need to be implemented and integrated.

The REMM, which is based on a unified framework, is one of the critical modules in the platform layer. It is used to control and manage all of the remote experiments. The REMM can deliver many services, such as the ability to conduct remote experiment, view experimental videos, manage and analyze experimental data, and edit the experimental tutorials, for students to support their collaborative online learning and co-creation with instructors. Meanwhile, it can also deliver services such as creating a new remote experiment, editing remote experiment, and posting the experiment tutorials, for instructors to support their teaching and co-creation with students. In the REMM, all of the remote experiments will be shown as web pages in the web browsers without requiring any software plug-ins.

The SAM is the other critical module in the platform layer used to organize, manage, and maintain all of the learning materials and the remote experiments delivered by the REMM. The SAM is the key module which supports the co-design of experiments and co-creation of knowledge between instructors and students. The SAM is also an online learning tool and can deliver many services, such as the ability to create new materials, edit materials, search materials and remote experiments, do homework online, and post to a discussion board, for students to support their learning experience. Meanwhile, the SAM enables instructional services, such as creating new pages, editing (adding and deleting) lectures, and posting homework, for instructors to support their teaching. In the SAM, all the materials also are shown as web pages in the web browsers. Users can create and edit the pages of SAM in markup language which allows the SAM to function as a cross-platform online learning tool.

Users can edit SAM content by following the user policy. The policy consists of management guidelines regarding the format of the web pages, the font and the size of words, the coherence of the terms to organize the content, and the quality of information. Operation requests are considered as user requests to perform certain tasks, such as search, edit, create, post, upload, and download. The selected content is in the client layer and supported by the platform layer. The Wiki-based RL platform integrates these implemented remote experiments to further address the different learning needs of students.

The integration of REMM and SAM delivers a student-centered collaborative and cooperative learning environment where students can put their learned skills to practice while improving their understanding of course concepts. By providing an outlet to address visual and kinesthetic learners through a web environment, the learning needs of students can be fulfilled. Based on the function and logic design for the platform layer, the functional modules are divided into the following two categories. (a) Database operation modules: these modules are used to deliver the APIs to support database operations (such as query, add, delete, modify, save, and view). (b) Development support modules: these modules are used to deliver the APIs to support development of other layers. The object-oriented programming (OOP) approach is used to define the classes and functions. Meanwhile, PHP language and JavaScript language are used to implement all of the classes and functions.

6.2.3 Database Layer

The core of the database layer is data resources development, which includes development and integration of a set of databases to support the operation of the platform layer. Particularly, three kinds of databases need to be developed and integrated – the study aids databases, the system databases, and the experiment management databases – in order to support the proposed Wiki-based remote laboratory platform. For the implementation of database layer, the most important task is to design and develop the DP. Some databases (including the search database, user information database, experiment database, page database, file database, and system parameters database) need to be developed and integrated into the DP. There are a total of 78 data tables implemented for the DP.

6.3 THE IMPLEMENTATION PROCESS OF THE WIKI-BASED RL PLATFORM

This process is used to combine the unified framework and MediaWiki together to generate the Wiki-based RL platform [177]. The system architecture of MediaWiki is shown in Figure 6.1. As the fundamental platform, the MediaWiki includes three layers: the UI layer, system logic layer, and data management layer. The web server of MediaWiki uses the Apache. The unified framework is integrated in three parts: (a) web server is changed to Apache and Node.js combined solution (in the red square), (b) remote experiment data is integrated into the database of the new Wiki-based RL platform (in the red square), and (c) components which are used to support remote experiments are integrated into the new Wiki-based RL platform (in the blue square). The implementation process of Wiki-based RL platform mainly includes two steps. In the first step, the fundamental environment, MediaWiki, is installed to provide the basics systems' APIs and development environment for the implementation of the next step. In the second step, the five modules and about 40 sub-modules will be implemented and integrated into the fundamental environment. Meanwhile, around 78 data tables will be implemented for the DP in this step as well.

The detailed systematic implementation process of the new RL platform includes three steps: setup, execution, and localization.

1. Setup: To set up the system, the features of MediaWiki need to be registered and configured. Before execution, the new platform needs to prepare the classes for loading all necessary resources like PHP, JavaScript, and CSS. The new platform is also directly built on top of MySQL database to store basic experiment data and system information. Some basic parameters of the unified framework for RL also need to be configured in this process, for example, IP configuration of experiment workstation and web camera and configuration of some system parameters of Apache web engine and Node.js web engine.

2. Execution: On the process of execution, Wiki markup typically contains a code that defines and implements custom XML tags, parser functions, and variables. The system will seamlessly integrate MediaWiki with its backing database. It also needs to check contents (such as remote experiments and learning materials) for integrity features and add functions for contents management. The scheduler and confliction management module is also executed to manage the remote experiments at this stage. With MediaWiki users' permission system, the security system for the remote experiment will be greatly improved.

3. Localization: MediaWiki has a localization engine to support developing localized systems which are adapted to various languages and regions without engineering changes. As a remote experiment is developed for different societies and countries, the internationalization and localization mechanisms are meant to adapting computer software for different languages, regional differences, and technical requirements of a target region.

6.4 SUMMARY

In this chapter, an innovative approach has been proposed for RL platform development based on a combination of advantages of Wiki technology and a unified framework. With this novel architecture, a more powerful online learning system supported by remote experiments is delivered. It offers a more flexible way to support students' collaborative learning.

Although the Wiki-based RL platform delivers new functions to support student-centered collaborative learning which cannot be supplied by the traditional RL platform, further development is still required to improve the novel platform's stability and usability. More specifically, issues that need improvement are as follows:

- More experiments need to be designed for different disciplines. With more and more new requirements from users, more and more new remote experiments need to be designed and implemented based on this novel platform to support student-centered collaborative and cooperative learning in future.
- New solutions and functions need to be explored and implemented. Currently, the experiments are controlled mainly via LtoN protocol based on the PC workstation in this novel platform. In the future, new functions will be integrated to support new hardware solutions to allow remote experiments based on Arduino Yun, Raspberry PI, Intel Galileo, etc. instead of PC workstation.

The Wiki-based RL platform is still version 1.0. So far, only a few pilot tests are conducted. Based on user feedback, several bugs in the software package have been fixed. More comprehensive testing will be done at other universities. Compared with the traditional RL platforms, the new platform can also provide more innovative user experience and give students more flexible selection to learn engineering knowledge without the limitation of time and space.

7

CASE STUDIES

7.1 INTRODUCTION

In this chapter, the detail design and implementation process of three remote experiments are presented. These three remote experiments are successfully implemented with different sub-frameworks and application architectures of the new-generation unified framework, and their user interface (UI) can be run on most of the popular web browsers without installing any additional software plug-in. To better guide the development of the remote laboratory (RL) systems, these remote experiments are discussed as the study cases. The capability of running remote experiments on terminal devices (such as desktops, laptops, and mobile devices) lets users gain insights by observing and interacting with the real instrument in a convenient way.

7.2 A REMOTE SHAPE MEMORY ALLOY (SMA) EXPERIMENT

The remote shape memory alloy (SMA) experiment is developed to demonstrate and analyze the characteristics of SMAs. It is used to study the hysteresis behavior of the wire actuator (SMA) and how the driving frequency changes the hysteresis loop. The experiment can apply a sinusoidal driving signal (AC) of 16.4 Volts in magnitude and, respectively, from 1/30 to 1/50 Hz in frequency to the SMA actuator. This remote experiment has been implemented in the University of Houston (UH) and Texas A&M University at Qatar (TAMUQ). Users can observe the wire movement through the webcam in the RL. They can record a number of cycles such as five for each test. Saved data are applied voltage, the displacement of the wire actuator, and time in seconds. The experimental data can be used for future analysis.

7.2.1 Methodology

An SMA has two phases: martensite and austenite. By increasing the temperature, an SMA will change from martensite to austenite. An SMA wire is easy to be lengthened in the phase of martensite at the lower temperature, and the length will return back to the original length in the phase of austenite at higher temperature. This is known as the shape memory effect (SME). The force generated by SMA wire in the phase of austenite is large, and the SME is controllable.

Consider the tracking control of SMA, and we use the equation from the implementation of nonlinear robust control on SMA. It uses a smooth compensator to control the displacement of SMA wire. The input is electrical current, and the output is the length change of SMA wire. Equations (7.1–7.3) are the mathematical models of the compensator:

$$e = y - y^d, \tag{7.1}$$

$$r = \dot{e} + \beta e, \text{ and} \tag{7.2}$$

$$u = u_f - Kr - \rho \tanh(ar). \tag{7.3}$$

The aim of the control system is to control the displacement of SMA length. y is the feedback of the system which is the real position of the SMA, and y^d is the desired position defined by the users. e is the error between the real position and the desired position. $\dot{e}$ is the derivative of the position error. u_f is a feedforward term to compensate the heat losses and provide a voltage to heat up the SMA. K and r expressed in Equation (7.2) are the proportional and derivative (PD) terms, and $\rho \tanh(ar)$ is the smooth robust compensator. The PD term in the feedback control increases the damping to reach the stable state of the system. The smooth robust compensator compensates the hysteresis of SMA, uncertainties in the system, and disturbances from the environment. u is the electrical current or the voltage applied to SMA to control its length change β, ρ, and a are the constants tuned in the experiment.

The feedback controller with the bang–bang compensator has expressions shown in Equation (7.4), where the feedforward term and the PD term are the same as those in the smooth compensator; and the difference is the bang–bang robust compensator, $\rho\text{sign}(r)$, defined by Equation (7.5):

$$u = u_f - Kr - \rho\text{sign}(r) \tag{7.4}$$

$$\text{sign}(r) \triangleq \begin{cases} 1, \text{if } r > 0 \\ -1, \text{if } r < 0 \end{cases}. \tag{7.5}$$

In the bang–bang compensator, when r is larger than zero, the gain of $\text{sign}(r)$ is positive; when r is less than zero, the gain of $\text{sign}(r)$ is negative. The r term defined in Equation (7.2) is dependent on the position error and the derivative of the position error.

The feedback controller with the saturation compensator has the expression shown in Equation (7.6), where the difference comparing to Equation (7.5) is the saturation term, $\rho\,\text{sat}\dfrac{r}{\varphi}$, defined by Equation (7.7):

$$u = u_f - Kr - \rho\text{sat}\left(\frac{r}{\varphi}\right) \tag{7.6}$$

$$\text{sat}\left(\frac{r}{\varphi}\right) \triangleq \begin{cases} 1, \text{if } r > \varphi \\ -1, \text{if } r < -\varphi \\ \dfrac{r}{\varphi}, \text{if } |r| < \varphi \end{cases}. \tag{7.7}$$

In the saturation compensator, r is defined by Equation (7.2), and φ is the constant defined by the user. The saturation compensator has a mathematical expression shown in Equation (7.7): when the variable r is larger than the defined constant φ, the gain of the saturation is positive; when the variable r is less than the negative constant φ the gain of the saturation is negative; and when the absolute value of the variable r is less than the constant φ, the saturation equals to the value of r divided by the constant φ, which is $\dfrac{r}{\varphi}$.

The bang–bang compensator has an instant change in magnitude from negative to positive; however, the saturation compensator has a linear increase in the magnitude change from negative to positive. The smooth compensator has more accurate results compared to the bang–bang compensator and saturation compensator. The smooth compensator follows the function, tanh(*ar*) in the change from a negative one to positive one. Thus, the smooth compensator is used in the SMA actuator in SMA device.

7.2.2 Experimental Implementation

The experiment consists of an SMA wire connected to the electronic circuitry that controls the voltage going to SMA wire. A displacement sensor is also attached to the SMA wire to sense the motion caused by the SMA during heating and cooling through the applied electric voltage. Remote SMA experiments are defined as real experiments conducted by the Internet users. These experiments use real instrumentation and components at a location different from where they are being controlled by the user.

7.2.2.1 Hardware Setup

As shown in Figure 7.1a, the SMA device is assembled by using both fabricated and off-the-shelf components. The main frame uses L-shaped aluminum bar fixed by screws. There are plexiglass sheets fixed on several sides of the SMA to protect and decorate the experiment. On the bottom, blue LED (light emitting diode) lights are used to light up the experiment. In the middle of SMA device, a red block is lifted by SMA wire and pulled down by two springs. The experiment consists of a workstation, NI DAQ 6008 USB, a DC power supply, a solid-state relay (SSR), a web camera, and SMA wires. A PC workstation runs Laboratory Virtual Instrument Engineering Workbench (LabVIEW) to control the voltage applied on the SMA wire. NI DAQ 6008 USB is a data acquisition device. It generates voltage output based on the signal from LabVIEW and senses voltage changes in the experiment to control the amount of voltage to SMA wire. The DC power supply is used to generate a consistent voltage on SMA. Pulse-width modulation (PWM) is used to control the voltage by using an SSR. The SSR can switch on and off swiftly to have a voltage value defined by the user even using a DC power supply. The camera is used to view the experimental change remotely. By applying voltage on SMA wire, the temperature of SMA will increase and the length of the wire will decrease.

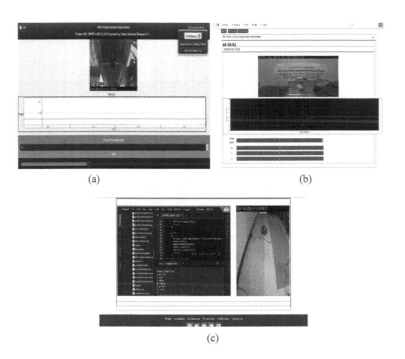

Figure 7.1 UI of three remote experiments: (a) SMA experiment, (b) PID motor speed control experiment, and (c) online programmable robot control experiment.

7.2.2.2 Software Implementation

The software implementation of remote SMA displacement experiment includes three tasks: the client module implementation, server module integration, and experimental control application implementation.

1. Client Web Module Implementation

   As shown in Figure 7.1a, to implement the web application, HTML5 Technology, which includes HyperText Markup Language (HTML), Cascading Style Sheets (CSS), and JQuery/JQuery-Mobile JavaScript libraries, is used to implement the UI. The socket.IO module is used to implement real-time data communication. With the server-based mash-up technology, the data is analyzed and reformatted on the remote server, and then the data is transmitted to the users' browsers. The UI includes three parts: (a) experimental real-time video, (b) real-time experimental data display, and (c) experimental control components.

2. Server Module Implementation

The server module is directly built on the top of a new assembled server engine scheme. It includes two server engines working together, Apache server engine and Node.js server engine. Based on the server-based mash-up technology used in the new framework, the Apache server engine is used to integrate the user interface widgets with web content and the real-time experimental video. Meanwhile, the Node.js server engine handles the experimental data in real-time transmission. Based on the online IM application architecture, a communication management module, which is used to handle real-time data transmission through the unique ID, is implemented. Moreover, the user management module, which is used to assign and manage the user unique ID, is implemented and integrated into the server module as well.

3. Integrating the New Version LabVIEW to Node.js (LtoN) Module into the Experimental Control Application

With the new LtoN module, students can conduct the experiment and save the experimental data to file system. More details of this new data transmission protocol are listed in the following:

• The new module includes two parts: the experimental equipment control part running in workstation and getting implemented by LabVIEW, and the server part running in web server which was developed using JavaScript language and enhanced by the Socket.IO.

• In this new module, we define our own special communication instruction set to implement real time experimental control commands and experimental data transmission.

• In the new module, some brief instructions to control experimental progress are designed to improve the data transmission performance.

Based on the above novel flexible framework, users can build up a distributed RL and rapidly and flexibly integrate any experiment developed by LabVIEW via its unique ID into the RL.

7.2.3 Results and Discussion

The remote SMA experiment has been tested by using the new RL system at TAMUQ. In the test, we performed a displacement control using a randomly chosen arbitrary value, which is 1.205 in, as the reference value. The reference switched between the arbitrary value and zero which was manually controlled in the graphical user interface (GUI). The data on the real displacement were recorded and directly downloaded from the GUI. In each test, the GUI recorded the data for 30 seconds, and the data was downloaded from the same GUI. The downloaded data for the real displacement and the displacement reference in each test was plotted in the same figure by using MATLAB.

7.3 A REMOTE PROPORTIONAL–DERIVATIVE–INTEGRAL (PID) MOTOR SPEED CONTROL EXPERIMENT

To better support the teaching of MECE 4327 (Mechanics, Controls, and Vibrations Laboratory) laboratory course, a remote proportional–derivative–integral (PID) motor speed control experiment was built to demonstrate the characteristics of proportional (P), proportional–integral (PI), proportional–derivative (PD), and PID controllers and visualize the process of remote tuning. It was established to help the study of the principles of dynamic systems analysis as well as to help achieve the system's optimum behavior depending on the applications. The controller was designed to control the angular velocity of a DC motor to follow various input signals (i.e., sinusoidal, triangular, or square waves) at specific frequencies varying from 0.1 to 1 Hz. Both the input and output angular velocities can be adjusted and displayed in real time, respectively. Meanwhile, the rotation of the DC motor can be observed through a webcam in the RL. It was recorded on the client station that the saved data included input angular velocity, measured output angular velocity, and time in seconds. With the Wiki-based RL platform, a PID motor speed control remote experiment was constructed and implemented at UH for an existing mechanical engineering lab course.

7.3.1 Methodology

Generally, PID controllers are widely used in industry, since they can be adjusted on site and have the capabilities of online automatic tuning [178]. The MECE 4327 laboratory course introduced the basic theory behind PID control. By tuning the three constants (K_p, K_i, K_d) in the PID controller algorithm, students will understand the concept of any PID application. The function of the PID algorithm is

$$u(t) = K_p e(t) + K_i \int_0^t e(t)\,dt + K_d \frac{de(t)}{dt}, \qquad (7.8)$$

where t denotes the time, $u(t)$ denotes the output of controller, K_p denotes the proportional gain, K_i denotes the integral gain, K_d denotes the derivative gain, and $e(t)$ denotes the error = setpoint – process variable.

P controller produces controller signal proportional to the current error value, which contributes to the bulk of the output change. I controller accumulates the errors from past, which eliminates the residual steady-state error. D controller can predict the system behavior, which decreases the setting time and improves the stability. Practically, tuning rules of Ziegler and Nichols are utilized to design each parameter of PID controller [179]. However, without hands-on experience of designing and tuning the parameters, students have difficulty in comprehending how each parameter contributes to PID controllers. To provide a remotely accessible platform for learning the characteristic of each parameter of PID controller, the remote PID controller experiments were designed to demonstrate the progress of designing PID controllers or modified PID controllers step by step, and to provide the visualization of real-time responses corresponding to changes in the parameters.

7.3.2 Experimental Implementation

7.3.2.1 Hardware Setup

The remote PID motor control hardware, as shown in Figure 7.1b, contains (a) DC motor with tachometer, (b) power amplifier, (c) power supply for the amplifier, (d) NI USB-6361 X Series DAQ, and (e) PC workstation. The NI USB-6361 X Series DAQ was utilized to both measure the voltage signal from tachometer as the rotation speed of DC motor and generate analog output voltage as control signal. The power amplifier and power supply provide the electronic power to drive the DC motor rotation.

All the functions in the bolded block are implemented by LabVIEW Virtual Instruments (VIs), which execute on the PC workstation. During the experiment, LabVIEW generates the reference signal, such as square wave and triangle wave at designated amplitude and frequency. Based on the reference signal and the feedback DC motor speed signal, the PID VIs calculate the control signal applied on the DC motor. To optimize the performance of the system, all the P, I, and D parameters could be tuned both locally and remotely.

7.3.2.2 Software Implementation

The Wiki-based RL platform framework includes three parts: client web application, server application, and experimental control application. The client web application is based on HTML, CSS, and JQuery/JQuery-Mobile JavaScript libraries. The mash-up technology is used for UI implementation. The client web application can be run in most of the current popular browsers such as Internet Explorer (IE), Firefox, Chrome, and Safari. The server application is based on web service technology and is directly built on top of MySQL database, Apache web server engine, and Node.js web server engine. The server application uses JSON and Socket.IO which is developed based on web socket protocol to implement real-time communication between the server application and the client web application. The server application runs on Centos Linux server. The experimental control application is based on the LabVIEW and uses Socket. IO for real-time communication with server application. The experimental control application runs on experimental control workstation which runs Window 7 OS. Our new unified framework for RL development is based on three vital technologies, which are the Socket.IO, Node-HTTP-Proxy, and HTTP Live Streaming (HLS) protocol. Node-HTTP-Proxy is used for transmission and traversing of experimental data and control commands through the firewall, and the novel video transmission approach is based on HLS protocol for real-time system monitoring. The real-time video streaming can be shown in most web browsers such as Firefox, Chrome, and Safari without any plug-in. However, a Java runtime environment is required for Microsoft IE. The mash-up technology is used for UI implementation.

In order to integrate the new remote PID controller experiment (further detail in Section 3) into the framework, three main tasks were accomplished: (1) GUI development, (2) system integration in server side, and (3) integrating LtoN module to the new PID controller experiment. Figure 7.1b shows the UI of the remote PID motor speed control experiment.

To better support the mobile learning (M-learning), based on the new mobile-optimized application architecture, the PID motor speed control experiment mobile application is designed and implemented. To implement the mobile-optimized remote experiment application, three key technologies, Ionic framework, Crosswalk platform, and Apache Cordova framework, have to be used. With Web 2.0 technology, Ionic framework uses the common codebase which includes HTML, CSS, and AngularJS/JavaScript. To enhance the mobile-optimized application performance, Crosswalk platform as the rendering engine is integrated into the mobile-optimized application. To automatically update the rendering engine based on the different platforms, the Crosswalk runs as a runtime engine in different mobile operating systems (OSs). In the implementation process, the common codebase is wrapped into the Cordova framework and rendered in the UI layer with the Crosswalk platform. Normally, the common codebase is reusable to deploy the mobile-optimized application to the different mobile platforms. The Crosswalk platform works as a middleware to connect the Ionic framework with the unified framework. With these three key technologies, all of the source code and software plugins are packaged with a new mobile-optimized remote experiment application. To package the new mobile-optimized application, Eclipse, a popular Integrated Development Environment (IDE), is used. The mobile-optimized remote PID motor speed control experiment can run on three different popular mobile Operation Systems (OSs).

7.3.3 Results and Discussion

In order to obtain student feedback regarding the effectiveness of the RL, a survey containing a series of questions was passed to the students. Of the 72 students enrolled in the course, 60 students performed the bonus RL experiment and completed the survey. The key finding is that most of the answers regarding the educational effectiveness center around "effective" (3 on a 1–4 scale) (questions 3–7). A majority of the students preferred hands-on experiments over RLs. This is expected since RLs, while still delivering the lab experience, are not exactly the same as going to the lab to perform the experiment in person (question 8). However, it is suspected that the convenience and intuitiveness of the RL also caused the students to nonetheless want to see more RLs in the future (question 9). Students were allowed to enter their own comments into the survey. The following are comments received in the surveys. User comments are listed as follows:

- The RL worked better and smoother on Firefox than Chrome and Safari.
- The procedure for saving data is little bit difficult; hope it improves for the better.
- Video is frozen, and I only see the data UI working well. I have to refresh the page.
- I could not zoom in on specific portions of the graphics in the RL.
- Maybe two hours is not enough, because the technical issues wasted more time.
- Sometimes, the system can produce my controlled value when my input value was zero.
- I signed up two times, but it wouldn't let me delete one of them.
- The video sometimes freezes, but I would have attempted further if I didn't have technical issues.

The participation was the 83% of the students who enrolled in this course. In all of the participants, 95% of the students were satisfied with accessibility and session time of the RL, while 32% of the students had some basic technical issues. The major reason is that they didn't read the user guide carefully and didn't follow the user guide step by step. Consequently, a video tutorial is necessary, and it should be more effective to assist students for RL in future.

In Figure 7.1, which shows the participants' answers to the survey questions, we can notice that more than 65% of the participants think this remote experiment is effective for their understanding basic PID theory. Especially, above 30% of the participants think that it is very effective. Although, around 30% of the participants still think that the remote experiment is not effective, we think that the positive response is sufficient to arrive at our goal and is a good start for integrating more remote experiments into future engineering courses.

Moreover, in the assessment, around 20% of the students mentioned technical problems in this assignment. From the comments received in the survey, we understand these problems drove them off. Meanwhile, the usability of this remote experiment requires improvement as well. These problems are going to be worked on, and it is hoped that at least some of the technical and usability obstacles would be removed. Moreover, the status of different students' networks had led to the real-time experimental video and experimental data graphics becoming frozen sometimes.

7.4 A PROGRAMMABLE REMOTE ROBOT CONTROL EXPERIMENT

Based on the new online programmable platform, the remote programmable robot control experiment has been implemented successfully. Users can implement and test their own robot control algorithm with C++ programming language with web-based UI anytime and anywhere. Meanwhile, the live experimental video can show the robot motion real-time following the users' designed control algorithm as the experiment result.

7.4.1 Methodology

To test the new online programmable platform, a simple robot control algorithm based on the common PID control model is developed. This algorithm is used to perform the basic position and motion control for an intelligent mobile robot named AmigoBot. The simplest method of formation control is a virtual structure approach, which lets (x_d, y_d) be the desired position of a robot. We define a virtual leader (x_{vl}, y_{vl}) for this robot having the same coordinate as the desired position (x_d, y_d) of the robot. The robot then tracks its virtual leader using PID control.

The kinematic equation of the mobile robot is

$$
\begin{bmatrix} \dot{r}_x \\ \dot{r}_y \\ \dot{\theta} \end{bmatrix} = \begin{bmatrix} v\cos\theta \\ v\sin\theta \\ \omega \end{bmatrix}.
\tag{7.9}
$$

Since there is no tangential velocity at the center of rotation of the robot, define another reference not in the center point of rotation:

$$
x = r_x + d\cos\theta
\tag{7.10}
$$

$$
y = r_y + d\cos\theta.
\tag{7.11}
$$

After deriving x and y once, the control variables υ and ω can be written directly in terms of $\dot{x}$ and $\dot{y}$:

$$
\begin{bmatrix} \upsilon \\ \omega \end{bmatrix} = \begin{bmatrix} \cos\theta & \sin\theta \\ -\dfrac{1}{d}\sin\theta & \dfrac{1}{d}\cos\theta \end{bmatrix} \begin{bmatrix} \dot{x} \\ \dot{y} \end{bmatrix}. \tag{7.12}
$$

Define $e_x = x_d - x$ and $e_y = y_d - y$; then a PID controller for (7.8–7.12) can be formulated as

$$
\dot{x} = K_p e_x + K_d \frac{de_x}{d_t} + K_i \int e_x d_t \cong K_p e_x + K_d \frac{\Delta e_x}{\Delta_t} + K_i \sum e_x \Delta_t \tag{7.13}
$$

$$
\dot{y} = K_p e_y + K_d \frac{de_y}{d_t} + K_i \int e_y d_t \cong K_p e_y + K_d \frac{\Delta e_y}{\Delta_t} + K_i \sum e_y \Delta_t. \tag{7.14}
$$

Based on this simple robot control algorithm, a robot's basic position and motion control program is developed to test the AmigoBot robot's actions. C++ language is adopted to implement the control program.

7.4.2 Experimental Implementation

7.4.2.1 Hardware Setup

In this experiment, AmigoBot robot is used for robots' motion control and planning, formation, flocking, source localization, swarm intelligence, and communication in robots' group. AmigoBot mobile robot has an Epia5000 mini-ITX board, which is a full mini-computer embedded in the robot. In this mini-computer, there is a serial port (for connecting to the robots), 4 USB ports with 2 USB bus (used to connect the USB wireless), 10/100 Ethernet port (for uploading and synchronization), and an 800 MHz C3 processor (equivalent to a P3 600 MHz). In this experiment, a WiBox 2100E wireless serial adapter, which is embedded in robot, is used to connect the robot to the Linux CentOS server through the D-Link wireless router. The experimental device is set up according to the intelligent client–server architecture. AmigoBot mobile robot is flexible to connect with the Linux CentOS server based on the HTTP protocol via wireless network. AmigoBot robot uses a WiBox wireless serial adapter to connect with Linux server through the DLink wireless router. One AXIS web camera is connected to the Linux server directly for real-time live experimental video. All experiment devices are set up in a wireless intranet. The Linux server works to manage all the low-level details of AmigoBot robots. The AmigoBot robot system, which works as the client, includes the sonar, motion control, positioning and movement control module, and sonar and motor encoder data collection. The hardware of remote programmable robot control experiment includes

- Robot (AmigoBot intelligent mobile robot)
- WiBox 2100E wireless serial adapter
- AXIS 2111M web camera
- DLink wireless router.

7.4.2.2 Software Implementation

The software implementation of the remote programmable robot control experiment includes three tasks, namely, the UI integration, platform layer integration, and implementation of experimental control application.

1. Client web application implementation

   The web application of the new remote programmable robot control experiment provides the users with a service to implement and verify their own robot control algorithm. As shown in Figure 7.1c, the web application based on the new programmable experiment platform includes three parts:
   - Real-time experimental video
   - Code input UI of the online web IDE
   - Running result shown on the UI of the online web IDE.

   From a technical point of view, the remote programmable robot experiment application uses Web 2.0 technology which includes HTML, CSS, and JQuery/JQuery-Mobile JavaScript libraries. Cloud9 provides the online web IDE UI. The real-time video is provided by the unified framework. With the server-based mash-up technology, the data was analyzed and reformatted on the remote server and then transmitted to the users' browsers.

2. Platform layer integration

   The new assembled server engine scheme and Cloud9 IDE are deployed and executed within Docker container, which is an open platform for developers to build, ship, and run distributed applications. Based on the Wiki-based RL framework, the platform layer is directly built on the top of a new assembled server engine scheme. It includes two server engines working together: Apache HTTP server engine and Node. js server engine. As the server-based mash-up technology used in the framework, the Apache HTTP server engine is applied to integrate the UI widgets with web content and the real-time experimental video. The Node.js server engine handles the experimental data in real-time transmission. Moreover, the Cloud9 is used to provide the online web IDE for control algorithm development. MySQL, a relational database management system, is used for data management.

3. Integration of the remote programmable robot control experiment

As shown in Figure 7.1c, the new remote programmable robot control experiment software part is built on Docker container. Two software modules are implemented in Docker container: Cloud9 IDE kernel and MediaWiki engine.

- The Cloud9 software package is the web-based IDE for the remote programmable experiment. Users can input and debug the algorithms in a real-time manner via web browsers.
- With the server-based mash-up technology, the web application interface is presented to users through the MediaWiki engine.
- Moreover, a real-time communication module supports real-time data transmission with mobile robot via Socket. IO protocol

7.4.3 Results and Discussion

In 2010, James Kuner at Google introduced the term "cloud robotics" to describe a new approach to robotics that takes advantage of the Internet as a resource for massively parallel computation and real-time sharing of vast data resources [170]. Based on this new remote programmable robot control experiment, a new development tool for cloud robotics applications is required to be implemented to provide a more powerful tool for cloud robotics and robot control research. In summary, the new online programmable platform offers a new online tool for the control algorithm and model development in engineering education and industries. It will be an innovative solution to benefit wireless control and sensing research of Industry 4.0.

7.5 SUMMARY

In this chapter, the successful development of three remote experiments based on the new-generation unified framework is discussed. These remote experiments include a Wiki-based remote PID motor speed control experiment for MECE 4327 course at UH, a remote SMA experiment with flexible framework, and an online programmable remote robot control experiment. End users can operate all of these remote experiments by using most of the web browsers on their terminal devices, which include desktops, laptops, and portable devices, without any other plug-in. In addition, this chapter offers users an online tool to promote active participation in lab practice as well. The capability of running a remote experiment on a terminal lets end users gain insights by observing and interacting with the real instrument in an efficient way. Compared with traditional approaches for remote experiment development, the new-generation unified framework offers a flexible approach to support the rapid implementation of users' own designed RL system.

8

CONCLUSIONS AND FUTURE WORKS

8.1 CONCLUSIONS

A new-generation unified framework for the development of a remote laboratory (RL) system is presented to solve five essential issues listed in Chapter 1. This new-generation unified framework includes five parts: a flexible framework for rapid integration with offline experimental equipment, a cloud-computing-based RLaaS-Frame (RLaaS: Remote Laboratory as a Service) framework for RL systems' rapid deployment, a mobile-optimized RL application architecture, an online programmable platform for remote programmable control experiment development, and a Wiki-based RL platform.

With the flexible framework established in Chapter 2, the essential issue, "How to design a flexible approach to rapidly integrate a new offline experimental equipment into an RL system?" is resolved. To resolve the issue, "How to rapidly deploy a stable RL system for widespread usage?" a cloud-computing-based RLaaS-Frame framework is presented in Chapter 3. In Chapter 4, a mobile-optimized RL application architecture is developed for resolving the issue, "How to design a mobile RL application that can run on different platforms easily with lower cost and less developmental effort?" An online programmable platform for the development of a remote programmable control experiment is presented in Chapter 5. It is mainly used to resolve the issue, "How to develop an online programmable platform to support the users to implement their own designed control algorithms and models to control the experimental equipment remotely in real time?" Finally, in Chapter 6, a Wiki-based RL platform is implemented for engineering education. These solutions form several holistic approaches to guide the rapid and efficient design and development of an RL system in future.

To better demonstrate the effectiveness of the new-generation unified framework, three remote experiments are discussed as the case studies. The detailed design and implementation processes of these three remote experiments are presented in Chapter 7. Compared to traditional approaches to RL system development, the new-generation unified framework offers a more flexible way to support users' RL system design and implementation, which will significantly benefit RL technology development in future.

8.2 FUTURE WORKS

Compared to the traditional approaches to RL system development, this new-generation unified framework has been significantly improved in whole process of RL system implementation. However, many tasks still need to be done in future, and they are listed below:

1. An approach for RL system reliability evaluation needs to be developed.

   The use of RL technology has increased in universities worldwide, and this technology has displayed advantages over the traditional hands-on laboratory. The performance and benefit of a RL could be affected by a number of reliability factors and their interactions in the System Development Life Cycle (SDLC). However, till now, almost all of RL research concerns either the technical aspects of RL system implementation or the learning-effect evaluation based on participants' opinions. Only a small portion of RL research efforts refer to the reliability factors and relevant impacts in the RL SDLC. As a consequence, most RL systems do not have effective reliability performance objectives or indicators, which results in the reduction in system lifetime, availability, and benefits and the increase in life cycle cost. To improve sustainable performance and benefits of RL system, the reliability performance objectives for RL system are selected based on the goal-setting theories. According to the results from the survey, key performance indicators (KPIs) are identified, and a conceptual model for KPIs is established to evaluate the reliability performance of the RL SDLC. Reliability performance objective and KPIs can be used to identify the success and weakness of the RL and are effective tools to manage RL SDLC performance for sustainable success and future improvement.

2. The AI technology needs to be integrated into this unified framework. Although the RL technology has shown to help engaged learning in the cognitive sense by providing a sense of realism not normally offered by e-learning, one major concern that is frequently raised is about the poor ability of students to perform experiments due to the lack of student assistance during their practical remote experiments of laboratory courses. With the continuous development of artificial intelligence (AI) technology and its extensive use, AI-technology-based intelligent tutorial system (ITS) is introduced to teaching the courses of different majors supported by e-learning (including mobile learning (M-learning)) opportunities. In addition, M-learning has demonstrated increasing impacts on online education; more and more mobile applications are designed and developed for M-learning. To assist student experiments and improve student concept mastery in-synch with laboratory course progress, an ITS-based RL framework for M-learning that will provide individualized learning based on iterative student assessments and assignments is proposed. By implementing a targeted, information-rich evaluation system on low-risk assignments for students, such as experimental tasks of laboratory courses, using an expert model based on artificial neural networks (ANNs), individual student needs can be addressed and future assignments and experimental tasks tailored based on the student's progression in future.

3. This unified framework needs to be integrated into Industry 4.0. Industry 4.0 is an emerging concept of automation and data exchange in industrial technologies, and it includes cyber-physical systems, the Internet of Things (IoT) technology, and cloud computing technology. With the IoT technology, cyber-physical systems communicate and cooperate with the industrial production equipment and with humans in real time. Both internal and cross-organizational services are offered and utilized by participants of the entire production chain. Based on the Industry 4.0 definition, the RL technology can be a suitable candidate to support wireless control and wireless sensing applications in Industry 4.0. This new-generation unified framework will provide a suitable tool for Industry 4.0 applications in future.

REFERENCES

1. M. Cooper, A. Donnelly, and J.M. Ferreira, "Remote controlled experiments for teaching over the Internet: A comparison of approaches developed in the PEARL project," In *Proceedings of the 19th Annual Conference of the Australian Society of Computers Education*, Auckland, Dec. 8–11, 2002.
2. Y.H. Elawady, and A.S. Tolba, "A general framework for remote laboratory access: A standardization point of view," In *Proceedings of the 2010 IEEE International Symposium on Signal Processing and Information Technology (ISSPIT)*, Luxor, pp. 485–490, Dec. 15–18, 2010.
3. M.G. Moore, and G. Kearsley, "Distance education: A systems view of online learning," *Cengage Learning*, Apr. 22, 2011.
4. N. Ertugrul, "Towards virtual laboratories: A survey of LabVIEW-based teaching/learning tools and future trends," *International Journal of Engineering Education (iJEE)*, vol. 16, no. 3, pp. 171–180, Jan. 1, 2000.
5. J. García-Zubia, P. Orduña, D. López-de-Ipiña, and G.R. Alves, "Addressing software impact in the design of remote laboratories," *IEEE Transactions on Industrial Electronics*, vol. 56, no. 12, pp. 4757–4767, Dec. 1, 2009.
6. M. Tawfik, D. Lowe, C. Salzmann, D. Gillet, E. Sancristobal, and M. Castro, "Defining the critical factors in the architectural design of remote laboratories," *IEEE-RITA*, vol. 10, no. 4, pp. 269–279. Nov. 1, 2015.
7. M.A. Prada, J.J. Fuertes, S. Alonso, S. García, and M. Domínguez, "Challenges and solutions in remote laboratories. Application to a remote laboratory of an electro-pneumatic classification cell," *Computers & Education*, vol. 85, pp. 180–190, Jul. 1, 2015.
8. M. Tawfik, C. Salzmann, D. Gillet, D. Lowe, H. Saliah-Hassane, E. Sancristobal, and M. Castro, "Laboratory as a service (LaaS): A novel paradigm for developing and implementing modular remote laboratories," *International Journal of Online Engineering (iJOE)*, vol. 10, no. 4, pp. 13–21, Oct. 1, 2014.

9. L. Tobarra, S. Ros, R. Pastor, R. Hernandez, M. Castro, A. Al-Zoubi, M. Dmour, A. Robles-Gomez, A. Caminero, and J. Cano, "Laboratories as a service integrated into learning management systems," In *Proceedings of the 13th IEEE International Conference on Remote Engineering and Virtual Instrumentation (REV)*, Madrid, pp. 97–102, Feb. 26–29, 2016.

10. S. Werner, A. Lauber, J. Becker, and E. Sax, "Cloud-based remote virtual prototyping platform for embedded control applications: Cloud-based infrastructure for large-scale embedded hardware-related programming laboratories," In *Proceedings of the 13th IEEE International Conference on Remote Engineering and Virtual Instrumentation (REV)*, Madrid, pp. 168–175, Feb. 26–29, 2016.

11. D. Gasevic, V. Kovanovic, S. Joksimovic, and G. Siemens, "Where is research on massive open online courses headed? A data analysis of the MOOC research initiative," *International Review of Research in Open and Distributed Learning*, vol. 15, no. 5, pp. 134–176, Oct. 3, 2014.

12. A. Juntunen, E. Jalonen, and S. Luukkainen, "Html 5 in mobile devices – Drivers and restraints," In *Processing of the 46th Hawaii International Conference on System Sciences (HICSS)*, IEEE, Hawaii, pp. 1053–1062, Jan. 7–10, 2013.

13. A. Charland, and B. Leroux, "Mobile application development: Web vs. native," *Communications of the ACM*, vol. 54, no. 5, pp. 49–53, May 1, 2011.

14. N. Serrano, J. Hernantes, and G. Gallardo, "Mobile web apps," *Software, IEEE*, vol. 30, no. 5, pp. 22–27, Sep. 1, 2013.

15. J. Chacón, M. Guinaldo, J. Sánchez, and S. Dormido. "A new generation of online laboratories for teaching automatic control," *IFAC-PapersOnLine*, vol. 48, no. 29, pp. 140–145, Jan. 1, 2015.

16. E. Besada-Portas, J. Bermúdez-Ortega, L. de la Torre, J.A. López-Orozco, and J.M. de la Cruz. "Lightweight node. js & EJsS-based web server for remote control laboratories," *IFAC-PapersOnLine*, vol. 49, no. 6, pp. 127–132, Dec. 31, 2016.

17. H. Temelta, M. Gokasan, and S. Bogosyan, "Hardware in the loop robot simulators for on-site and remote education in robotics," *International Journal of Engineering Education*, vol. 22, no. 4, p. 815, Aug. 1, 2006.

18. W. Hu, G.P. Liu, D. Rees, and Y. Qiao, "Design and implementation of web-based control laboratory for test rigs in geographically diverse locations," *IEEE Transactions on Industrial Electronics*, vol. 55, no. 6, pp. 2343–2354, Jun. 1, 2008.

19. L. Gomes, and S. Bosgoyan, "Current trends in remote laboratories," *IEEE. Trans. on Industrial Electronics*, vol. 56, no. 12, pp. 4744–4756, Dec. 1, 2009.

20. J. Rodriguez-Andina, L. Gomes, and S. Bogosyan, "Current trends in industrial electronics education," *IEEE Transactions on Industrial Electronics*, vol. 57, no. 10, pp. 3242–3244, Oct. 1, 2010.

21. N. Wang, M. Ho, Q. Lan, X. Chen, G. Song, and H. Parsaei, "Developing a remote laboratory at TAMUQ based on a novel unified framework," In *Processing of the 122rd ASEE Annual Conference and Exposition*, age, 26, p. 1, Nov. 11–13, 2015.

22. N. Wang, X. Chen, G. Song, and H. Parsaei, "An experiment scheduler and federated authentication solution for remote laboratory access," *International Journal of Online Engineering (iJOE)*, vol. 11, no. 3, Jun. 1, 2015.

23. N. Wang, X. Chen, G. Song, and H. Parsaei, "Using node-HTTP-proxy for remote experiment data transmission traversing firewall," *International Journal of Online Engineering (iJOE)*, vol. 11, no. 2, pp. 4–9, Mar. 1, 2015.

24. N. Wang, X. Chen, G. Song, and H. Parsaei, "A novel real-time video transmission approach for remote laboratory development," *International Journal of Online Engineering (iJOE)*, vol. 11, no. 2, pp. 4–9, Mar. 1, 2015.

25. N. Wang, X. Chen, G. Song, Q. Lan, and H.R. Parsaei, "Design of a New Mobile-Optimized Remote Laboratory Application Architecture for M-Learning," *IEEE Transactions on Industrial Electronics*, vol. 64, no. 3, pp. 2382–2391, Mar. 1, 2017.

26. N. Wang, X. Chen, Q. Lan, G. Song, H. Parsaei, and S.C. Ho, "A novel wiki-based remote laboratory platform for engineering education," *IEEE Transactions on Learning Technologies*, vol. 10, no. 3, pp. 331–341, Jul. 20, 2016.

27. J. Zhu, N. Wang, Q. Lan, G. Song, X. Chen, and H. Parsaei, "Develop a remote PID motor control experiment for engineering technology education–A case study," *2016 Earth and Space*, Feb. 16–18, pp. 1016.

28. N. Wang, J. Weng, X. Chen, G. Song, and H. Parsaei, "Development of a remote shape memory alloy experiment for engineering education," *Engineering Education Letters*, vol. 2, pp. 1–20, Apr. 23, 2015.

29. N. Wang, J. Weng, M. Ho, X. Chen, G. Song, and H. Parsaei, "Development of a remote Sma experiment-A case study," In *Qatar Foundation Annual Research Conference*, no. 1, p. ITPP0944, Nov. 11, 2014.

30. N. Rosenberg, and R.R. Nelson, "American universities and technical advance in industry," *Research Policy*, vol. 23, no. 3, pp. 323–348, May 1, 1994.

31. K. Hinkelmann, ed., "Design and analysis of experiments, special designs and applications," John Wiley & Sons, Hoboken, NJ, *vol. 3, Dec. 28, 2011.*

32. J.E.V. Aken, "Management research based on the paradigm of the design sciences: the quest for field-tested and grounded technological rules," *Journal of Management Studies*, vol. 41, no. 2, pp. 219–246, Mar. 1, 2004.

33. P. Moreno-Ger, J. Torrente, J. Bustamante, C. Fernández-Galaz, B. Fernández-Manjón, and M.D. Comas-Rengifo, "Application of a low-cost web-based simulation to improve students' practical skills in medical education," *International Journal of Medical Informatics*, vol. 79, no. 6, pp. 459–467, Jan. 30, 2010.

34. V. Potkonjak, M. Gardner, V. Callaghan, P. Mattila, C. Guetl, V.M. Petrović, and K. Jovanović, "Virtual laboratories for education in science, technology, and engineering: A review," *Computers & Education*, vol. 95, pp. 309–327, Apr. 30, 2016.

35. M. Cooper, and J.M. Ferreira, "Remote laboratories extending access to science and engineering curricular," *IEEE Transactions on Learning Technologies*, vol. 2, no. 4, pp. 342–353, Oct. 1, 2009.

36. S. You, T. Wang, R. Eagleson, C. Meng, and Q. Zhang, "A low-cost internet-based tele-robotic system for access to remote laboratories," *Artificial Intelligence in Engineering*, vol. 15, no. 3, pp. 265–279, Jul. 31, 2001.

37. A. Melkonyan, A. Gampe, M. Pontual, G. Huang, and D. Akopian, "Facilitating remote laboratory deployments using a relay gateway server architecture," *IEEE Transactions on Industrial Electronics*, vol. 61, no. 1, pp. 477–485, Jan. 1, 2014.

38. I. Santana, M. Ferre, E. Izaguirre, R. Aracil, and L. Hernandez, "Remote laboratories for education and research purposes in automatic control systems," *IEEE Transactions on Industrial Informatics*, vol. 9, no. 1, pp. 547–556, Feb. 1, 2013.

39. L. Harasim, "A history of e-learning: Shift happened," In *The international handbook of virtual learning environments*, J. Weiss, J. Nolan, J. Hunsinger, and P. Trifonas (eds.), pp. 59–94, *Springer,* Dordrecht, *Jan.* 1, 2006.

40. A.S. Nada, F.A. Alzahrani, and O.B. Abouelatta, "Interactive web-based virtual electrical lab," *Journal of American Science*, vol. 8, no. 11, pp. 475–484, Aug. 1, 2012.

41. L.C. Gomes de Freitas, M.G. Simões, C.A. Canesin, and L.C. De Freitas, "Performance evaluation of a novel hybrid multipulse rectifier for utility interface of power electronic converters," *IEEE Transactions on Industrial Electronics*, vol. 54, no. 6, pp. 3030–3041, Dec. 1, 2007.

42. A.A. Kist, P. Gibbings, A.D. Maxwell, and H. Jolly, "Supporting remote laboratory activities at an institutional level," *International Journal of Online Engineering*, vol. 9, pp. 38–47, Jun. 2, 2013.

43. M. Abdulwahed, and Z.K. Nagy, "Applying Kolb's experiential learning cycle for laboratory education," *Journal of Engineering Education*, vol. 98, no. 3, pp. 283–294, Jul. 1, 2009.
44. A. Barrios, S. Panche, M. Duque, V.H. Grisales, F. Prieto, J.L. Villa, P. Chevrel, and M. Canu, "A multi-user remote academic laboratory system," *Computers & Education*, vol. 62, pp. 111–122, Mar. 31, 2013.
45. V.J. Harward, J.A., Del Alamo, S.R. Lerman, P.H. Bailey, J. Carpenter, K. DeLong, C. Felknor, J. Hardison, B. Harrison, I. Jabbour, and P.D. Long, "The ilab shared architecture: A web services infrastructure to build communities of internet accessible laboratories," *Proceedings of the IEEE*, vol. 96, no. 6, pp. 931–950, Jun. 1, 2008.
46. S. Colbran, and M. Schulz, "An update to the software architecture of the iLab service broker," In *Proceeding of the IEEE International Conference REV Instrumentation*, Bangkok, pp. 90–93, Feb. 26–29, 2015.
47. L.A. Mendes, L. Li, P.H. Bailey, K.R. DeLong, and J.A. del Alamo, "Experiment lab server architecture: A web services approach to supporting interactive LabVIEW-based remote experiments under MIT's iLab shared architecture," In *Proceedings of the IEEE International Conference REV Instrumentation*, Madrid, pp. 293–305, Feb. 24–26, 2016.
48. P. Orduña, J. Irurzun, L. Rodriguez-Gil, J.G. Zubía, F. Gazzola, and D. López-de-Ipiña, "Adding new features to new and existing remote experiments through their integration in WebLab-Deusto," *International Journal of Online Engineering (iJOE)*, vol. 7, no. S2, pp. 33–39, Oct. 1, 2011.
49. P. Orduña, A. Gómez-Goiri, L. Rodriguez-Gil, J. Diego, D. López-de-Ipiña, and J. Garcia-Zubia, "wCloud: Automatic generation of WebLab-Deusto deployments in the Cloud," In *Proceedings of the IEEE International Conference REV Instrumentation*, Bangkok, Thailand, pp. 223–229, Feb. 24–26, 2015.
50. I. Gustavsson, K. Nilsson, J. Zackrisson, J. Garcia-Zubia, U. Hernandez-Jayo, A. Nafalski, Z. Nedic, O. Gol, J. Machotka, M.I. Pettersson, and T. Lago, "On objectives of instructional laboratories, individual assessment, and use of collaborative remote laboratories," *IEEE Transactions on Learning Technologies*, vol. 2, no. 4, pp. 263–274, Oct. 1, 2009.
51. M. Tawfik, E. Sancristobal, S. Martin, R. Gil, G. Diaz, A. Colmenar, J. Peire, M. Castro, K. Nilsson, J. Zackrisson, and L. Hakansson, "Virtual instrument systems in reality (VISIR) for remote wiring and measurement of electronic circuits on breadboard," *IEEE Transactions on Learning Technologies*, vol. 6, no. 1, pp. 60–72, Jan. 1, 2013.

52. A. Macho, E. Sancristobal, M. Rodr, and M. Castro, "Remote laboratories for electronics and new steps in learning process integration," In *Proceedings of the IEEE International Conference REV Instrumentation*, Madrid, pp. 112–117, Feb. 24–26, 2016.

53. D. Lowe, S. Murray, E. Lindsay, and D. Liu, "Evolving remote laboratory architectures to leverage emerging internet technologies," *IEEE Transactions on Learning Technologies*, vol. 2, no. 4, pp. 289–294, Oct. 1, 2009.

54. D. Lowe, T. Machet, and T. Kostulski, "UTS remote labs, labshare, Sahara architecture," In *Using Remote Labs in Education*, Javier Garcia Zubia and Gustavo R. Alves (eds.), Bilbao, Spain: University of Deusto, pp. 403–424, Feb. 1, 2012.

55. T. Richter, D. Boehringer, and S. Jeschke, "Lila: A european project on networked experiments," In *Automation, Communication and Cybernetics in Science and Engineering*, pp. 307–317, Jan. 1, 2011.

56. T. Richter, Y. Tetour, and D. Boehringer, "Library of labs – A European project on the dissemination of remote experiments and virtual laboratories," In *Proceedings ISM Conference*, Dana Point, CA, pp. 543–548, Dec. 5–8, 2011.

57. V. Mateos, A. Gallardo, T. Richter, L. Bellido, P. Debicki, and V. Villagrá, "Lila booking system: Architecture and conceptual model of a rig booking system for on-line laboratories," *International Journal of Online Engineering (iJOE)*, vol. 7, no. 4, pp. 26–35, Nov. 10, 2012.

58. C. Olmi, X. Chen, and G. Song, "A framework for developing scalable remote experiment laboratory," In *E-Learn: World Conference on E-Learning in Corporate, Government, Healthcare, and Higher Education*, Honolulu, HI, pp. 2045–2050, Oct. 18–21, 2011.

59. X. Chen, D. Osakue, N. Wang, H. Parsaei, and G. Song, "Development of a remote experiment under a unified remote laboratory framework," *QScience Proceedings*, Doha, pp. 7–13, July. 1, 2014.

60. S. Li, J. Huai, and B. Bhargava, "Building high performance communication services for digital libraries," In *Proceeding of the Forum on Research and Technology Advances in Digital Libraries*, McLean, VA, pp. 34–39, May 1, 1995.

61. D. Gillet, C. Salzmann, R. Longchamp, and D. Bonvin, "Telepresence: An opportunity to develop real-world experimentation in education," In *Processing of the European Control Conference (ECC), IEEE*, Brussels, pp. 1646–1651, Jul. 1–4, 1997.

62. H. Vargas, J. Sánchez, N. Duro, R. Dormido, S. Dormido-Canto, G. Farias, S. Dormido, F. Esquembre, C. Salzmann, and D. Gillet, "A systematic two-layer approach to develop web-based experimentation environments for control engineering education," *Intelligent Automation & Soft Computing*, vol. 14, no. 4, pp. 505–524, Jan. 1, 2008.

63. N. Duro, R. Dormido, H. Vargas, S. Dormido-Canto, J. Sánchez, G. Farias, S. Dormido, and F. Esquembre, "An integrated virtual and remote control lab: The three-tank system as a case study," *Computing in Science & Engineering*, vol. 10, no. 4, pp. 50–59, Jul. 1, 2008.

64. M.J. Callaghan, J. Harkin, E. McColgan, T.M. McGinnity, and L.P. Maguire, "Client–server architecture for collaborative remote experimentation," *Journal of Network and Computer Applications*, vol. 30, no. 4, pp. 1295–1308, 2007.

65. C. Gravier, J. Fayolle, B. Bayard, M. Ates, and J. Lardon, "State of the art about remote laboratories paradigms – Foundations of ongoing mutations," *International Journal of Online Engineering(iJOE)*, vol. 4, no. 1, pp. 19–25, Feb. 1, 2008.

66. M.T. Restivo, J. Mendes, A.M. Lopes, C.M. Silva, and F. Chouzal, "A remote laboratory in engineering measurement," *IEEE Transactions on Industrial Electronics*, vol. 56, pp. 4836–4843, Dec. 1, 2009.

67. R. Pastor, J. Sanchez, and S. Dormido, "Web-based virtual lab and remote experimentation using easy java simulations," In *Processing of the 16th IFAC World Congress,* Prague, Vol. 16, Part 1, pp. 2289, Dec. 31, 2005.

68. P. Bistak, "Remote laboratory server based on java Matlab interface," In *Processing of the 14th International Conference on Interactive Collaborative Learning (ICL 2011),* Piestany, pp. 344–347, Sep. 21–24, 2011.

69. P. Bailey, and J. Hardison, "iLab batched experiment architecture: Client and lab server design," In *Processing of the MIT iCampus iLabs Software Architecture Workshop,* Hardison, NC, June 13–15, 2006.

70. S. Jeschke, T. Richter, and U. Sinha, "Embedding virtual and remote experiments into a cooperative knowledge space," In *Paper presented at the 38th ASEE/IEEE Frontiers in Education Conference,* Saratoga Springs, NY, Oct. 22–25, 2008.

71. L.I. Brodsky, V.V. Ivanov, Y.L. Kalaydzidis, A.M. Leontovich, V.K. Nikolaev, S.I. Feranchuk, and V.A. Drachev, "GeneBee-NET: Internet-based server for analyzing biopolymers structure," *Biochemistry-New York-English Translation of Biokhimiya*, vol. 60, no. 8, pp. 923–928, Aug. 1, 1995.

72. P. Mell, and T. Grance, "The NIST definition of cloud computing," *National iInstitute of standards and technology*, Special publication 800-145, Gaithersburg, MD, Jan. 1, 2011.
73. B. Sosinsky, "Defining Cloud Computing," In *Cloud Computing Bible*, John Wiley & Sons, Hoboken, NJ, pp. 1–22, Jan. 1, 2010.
74. Q. Zhang, L. Cheng, and R. Boutaba, "Cloud computing: state-of-the-art and research challenges," *Journal of Internet Services and Applications*, vol. 1, no. 1, pp. 7–18, May 1, 2010.
75. D.E. Williams, "Virtualization with Xen (tm): Including XenEnterprise, XenServer, and XenExpress," *Syngress*, Burlington, MA, Jul. 3, 2007.
76. G. Juve, E. Deelman, K. Vahi, G. Mehta, B. Berriman, B.P. Berman, and P. Maechling, "Scientific workflow applications on Amazon EC2," In *Processing of the 5th IEEE International Conference on E-Science Workshops*, Oxford, pp. 59–66, Dec. 9, 2009.
77. B. Wilder, "Cloud architecture patterns: Using microsoft azure," *O'Reilly Media, Inc.*, Sep. 20, 2012.
78. S. Srivastava, V. Trehan, P. Yadav, N. Manga, and S. Gupta, "Google app engine," *International Journal of Engineering and Innovative Technology (IJEIT)*, vol. 2, no. 3, pp. 163–165, Feb. 1, 2012.
79. S.P.T. Krishnan, and J.L.U. Gonzalez, "Google app engine," In *Building Your Next Big Thing with Google Cloud Platform*, Apress, Berkeley, CA, pp. 83–122, 2015.
80. D. Huang, "Mobile cloud computing," *IEEE COMSOC Multimedia Communications Technical Committee (MMTC) E-Letter*, vol. 6, no. 10, pp. 27–31, Oct. 6–9, 2011.
81. H.T. Dinh, C. Lee, D. Niyato, and P. Wang, "A survey of mobile cloud computing: Architecture, applications, and approaches," *Wireless Communications and Mobile Computing*, vol. 13, no. 18, pp. 1587–1611, Dec. 25, 2013.
82. H. Qi, and A. Gani, "Research on mobile cloud computing: Review, trend and perspectives," In *Processing of the Second International Conference on Digital Information and Communication Technology and It's Applications (DICTAP), IEEE*, Bangkok, pp. 195–202, May 16–19, 2012.
83. iLab Solutions, www.ilabsolutions.com, Jul. 16, 2017.
84. M. Tawfik, C. Salzmann, D. Gillet, D. Lowe, H. Saliah-Hassane, E. Sancristobal, and M. Castro, "Laboratory as a Service (LaaS): A model for developing and implementing remote laboratories as modular components," In *Processing of the 11th International Conference on Remote Engineering and Virtual Instrumentation (REV), IEEE*, Porto, pp. 11–20, Feb. 26–29, 2014.

85. A. Longo, M. Zappatore, and M.A. Bochicchio, "Towards massive open online laboratories: An experience about electromagnetic crowd sensing," In *Processing of the 12th International Conference on Remote Engineering and Virtual Instrumentation (REV), IEEE,* Bangkok, pp. 43–51, Feb. 26–29, 2015.
86. C. Salzmann, D. Gillet, and Y. Piguet, "MOOLs for MOOCs: A first edX scalable implementation," In *Processing of the 13th International Conference on Remote Engineering and Virtual Instrumentation (REV), IEEE,* Madrid, pp. 246–251, Feb. 26–29, 2016.
87. D. May, C. Terkowsky, T. Haertel, and C. Pleul, "Bringing remote labs and mobile learning together," *International Journal of Interactive Mobile Technologies (iJIM),* vol. 7, no. 3, pp. 54–62, Jun. 26, 2013.
88. J.B. Silva, W. Rochadel, J.P. Simão, R. Marcelino, and V. Gruber, "Using mobile remote experimentation to teach physics in public school," In *International Computer Aided Blended Learning Conference,* Florianópolis, Brazil, Nov. 6–8, 2013.
89. S.A. Saleh, and B.S. Ahmad, "Mobile learning: A systematic review," *International Journal of Computer Applications (IJOCA),* vol. 114, no. 11, pp. 1–5, Jan. 1, 2015.
90. H. Crompton, D. Burke, K.H. Gregory, and C. Gräbe, "The use of mobile learning in science: A systematic review," *Journal of Science Education and Technology,* vol. 25, no. 2, pp. 149–160, Apr. 1, 2016.
91. D. Ivanc, R. Vasiu, and M. Onita, "Usability evaluation of a LMS mobile web interface," In *Processing of the 2012 International Conference on Information and Software Technologies,* Springer Berlin Heidelberg, pp. 348–361, Sep. 13–16, 2012.
92. S. Teri, A. Acai, D. Griffith, Q. Mahmoud, D.W. Ma, and G. Newton, "Student use and pedagogical impact of a mobile learning application," *Biochemistry and Molecular Biology Education,* vol. 42, no. 2, pp. 121–135, Mar. 1, 2014.
93. B. Deaky, D.G. Zutin, and P.H. Bailey, "The first android client application for the iLab shared architecture," *International Journal of Online Engineering (iJOE),* vol. 8, no. 1, pp. 4–7, Feb. 2, 2012.
94. A. Hossain, J. Canning, S. Ast, P.J. Rutledge, T.L. Yen, and A. Jamalipour, "Lab-in-a-phone: Smartphone-based portable fluorometer for pH measurements of environmental water," *Sensors Journal, IEEE,* vol. 15, no. 9, pp. 5095–5102, Sep. 1, 2015.
95. A. Alkouz, A.Y. Al-Zoubi, and O. Mohammed, "J2ME-based mobile virtual laboratory for engineering education," *International Journal of Interactive Mobile Technologies (iJIM),* vol. 2, no. 2, pp. 5–10, Apr. 1, 2008.

96. D. Chaos, J. Chacón, J.A. Lopez-Orozco, and S. Dormido, "Virtual and remote robotic laboratory using EJS, MATLAB and LabVIEW," *Sensors*, vol. 13, no. 2, pp. 2595–2612, Feb. 21, 2013.

97. C.A. Garc, and P. Merino, "Remote control and instrumentation of Android devices," In *2016 13th International Conference on Remote Engineering and Virtual Instrumentation (REV)*, Madrid, pp. 190–195, Feb. 24–27, 2016.

98. C. Onime, and O. Abiona, "3D mobile augmented reality interface for laboratory experiments," *International Journal of Communications, Network and System Sciences*, vol. 9, no. 4, pp. 67–76, Apr. 27, 2016.

99. J. Garcia-Zubia, D. López-de-Ipiña, and P. Orduña, "Mobile devices and remote labs in engineering education," In *2008 Eighth IEEE International Conference on Advanced Learning Technologies*, Santander, pp. 620–622, Jul. 1, 2008.

100. C. Terkowsky, C. Pleul, I. Jahnke, and A.E. Tekkaya, "Tele-operated laboratories for online production engineering education-platform for e-learning and telemetric experimentation (PeTEX)," *International Journal of Online Engineering (iJOE)*, vol. 7, no. S1, pp. 37–43, Sep. 2, 2011.

101. J.B. Silva, W. Rochadel, J.P.S. Simao, S.M.S. Bilessimo, and P.C. Nicolete, "Using mobile devices for conducting experimental practices in basic education," *Maria Teresa Restivo; Alberto Cardoso; António Mendes Lopes. (Org.). Online Experimentation: Emerging Technologies and IoT. 1ed.Barcelona - Espanha: IFSA Publishing*, vol. 1, pp. 401–417, Feb. 2016.

102. J.B. Ortega, E.B. Portas, J.A.L. Orozco, J.A.B. Seco, and J.M. Cruz, "Remote web-based control laboratory for mobile devices based on EJsS, Raspberry Pi and Node.js," *IFAC-Papers OnLine*, vol. 48, no. 29, pp. 158–163, Jan. 1, 2015.

103. F. Lustig, J. Dvorak, and P. Brom, "Simple modular system "iSES Remote Lab SDK" for creation of remote experiments accessible from PC, tablets and mobile phones: Workshop," In *2016 13th International Conference on Remote Engineering and Virtual Instrumentation (REV)*, Madrid, pp. 406–408. Feb. 26–29. 2016.

104. J. Hylén, "Open educational resources: Opportunities and challenges," *Proceedings of Open Education*, Logan, UT, pp. 49–63, Sep. 27, 2006.

105. N. Butcher, "A basic guide to open educational resources (OER)," *Commonwealth of Learning (COL)*, Oct. 2015.

106. M. Parameswaran, and A.B. Whinston, "Social computing: An overview," *Communications of the Association for Information Systems*, vol. 19, no. 1, p. 37, Jun. 1, 2007.

107. M. Parameswaran, and A.B., Whinston, "Research issues in social computing," *Journal of the Association for Information Systems*, vol. 8, no. 6, p. 336, Jul. 1, 2007.

108. M. Tavakolifard, and K.C. Almeroth, "Social computing: An intersection of recommender systems, trust/reputation systems, and social networks," *IEEE Network*, vol. 26, no. 4, Jul. 1, 2012.

109. S. Mader, "Wikipatterns," *John Wiley & Sons*, Hoboken, NJ, Oct. 2008.

110. D.J. Barrett, "What's MediaWiki?," *MediaWiki, Sebastropol: O'Reilly Media*, ISBN:978-0-596-51979-7, pp. 4–5, Jan. 2008.

111. A. Pilav-Velic, and A. Habul, "Does social computing provide satisfaction for modern learner?," In *Proceedings of the ITI 2011 33rd International Conference on Information Technology Interfaces (ITI)*, Dubrovnik, Croatia, pp. 319–324, Jun 27–29, 2011.

112. C. Redecker, K. Ala-Mutka, and Y. Punie, "Learning 2.0 – The impact of social media on learning in Europe," *Policy Brief. JRC Scientific and Technical Report*. EUR JRC56958 EN, available from http://bit. ly/cljlpq [Accessed 6th February 2011], Jan. 2011.

113. C. Redecker, A. Haché, and C. Centeno, "Using information and communication technologies to promote education and employment opportunities for immigrants and ethnic minorities," *Joint Research Centre, European Commission,* May 1, 2010.

114. R.M. Felder, G.N. Felder, and E.J. Dietz, "A longitudinal study of engineering student performance and retention. V. Comparisons with traditionally-taught students," *Journal of Engineering Education*, vol. 87, no. 4, pp. 469–480, Oct. 1, 1998.

115. R.M. Felder, and R. Brent, "Understanding student differences," *Journal of Engineering Education*, vol. 94, no. 1, pp. 57–72, Jan. 1, 2005.

116. N. Wang, X. Chen, G. Song, and H. Parsaei, "Remote experiment development using an improved unified framework," In *World Conference on E-Learning in Corporate, Government, Healthcare, and Higher Education,* vol. 2014, no. 1, pp. 2003–2010, Oct. 20, 2014.

117. A. Maiti, A.A. Kist, and A.D. Maxwell, "Real-time remote access laboratory with distributed and modular design," *IEEE Transactions on Industrial Electronics*, vol. 62, no. 6, pp. 3607–3618, Jun. 1, 2015.

118. M. Kaluz, J. Garcia-Zubia, M. Fikar, and Ľ. Čirka, "A flexible and configurable architecture for automatic control remote laboratories," *IEEE Transactions on Learning Technologies*, vol. 8, no. 3, pp. 299–310, Jul. 1, 2015.

119. S. Farah, A. Benachenhou, G. Neveux, D. Barataud, G. Andrieu, and T. Fredon, "Multi-user and real-time flexible remote laboratory architecture for collaborative and cooperative pedagogical scenarios," *iJOE*, vol. 12, no. 4, pp. 33–36, Aug. 1, 2016.

120. D. Lowe, "MOOLs: Massive open online laboratories: An analysis of scale and feasibility," In *Proceedings IEEE International Conference REV Instrumentation*, Porto, pp. 1–6, Feb. 26–29, 2014.

121. L. Benetazzo, M. Bertocco, F. Ferraris, A. Ferrero, C. Offelli, M. Parvis, and V. Piuri, "A Web-based distributed virtual educational laboratory," *IEEE Transactions on Instrumentation and Measurement*, vol. 49, no. 2, pp. 349–356, Apr. 1, 2000.

122. M. Auer, A. Pester, D. Ursutiu, and C. Samoila, "Distributed virtual and remote labs in engineering," In *Proceedings of the IEEE International Conference on Industrial Technology*, Maribor, vol. 2, pp. 1208–1213, Dec. 30, 2003.

123. E. Bogdanov, C. Salzmann, and D. Gillet, "Widget-based approach for remote control labs," *IFAC Proceedings*, vol. 45, no. 11, pp. 189–193, Jan. 1, 2012.

124. D. Gillet, "Personal learning environments as enablers for connectivist MOOCs," In *Proceedings on 12th ITBHET International Conference*, Antalya, pp. 1–5, Oct. 10, 2013.

125. C.T. Moncreiff, "Computer network chat room based on channel broadcast in real time," *U.S. Patent 6,061,716*, 2000.

126. R. Peris, M.A. Gimeno, D. Pinazo, G. Ortet, V. Carrero, M. Sanchiz, and I. Ibanez, "Online chat rooms: Virtual spaces of interaction for socially oriented people," *CyberPsychology & Behavior*, vol. 5, no. 1, pp. 43–51, Feb. 1, 2002.

127. Z. Yang, and Q. Liu, "Research and development of web-based virtual online classroom," *Computers & Education*, vol. 48, no. 2, pp. 171–184, Feb. 28, 2007.

128. M. Dereshiwsky, "Equity in the online classroom: Adolescent to adult," In *Proceedings Social Justice Instruction Conference*, pp. 33–42, 2016.

129. S. Tilkov, and S. Vinoski, "Node.js: Using JavaScript to build high-performance network programs," *IEEE Internet Computing*, vol. 14, no. 6, p. 80, Nov. 1, 2010.

130. K. Lei, Y. Ma, and Z. Tan, "Performance comparison and evaluation of web development technologies in php, python, and node.js," In *Proceedings 17th IEEE CSE International Conference*, Chengdu, pp. 661–668, Dec. 19, 2014.

131. S.L. Bangare, S. Gupta, M. Dalal, and A. Inamdar, "Using node. js to build high speed and scalable backend database server," In *Proceedings NCPCI Conference*, vol. 2016, pp. 19, Mar. 2016.

132. T. Hughes-Croucher, and M. Wilson, "Up and running with Node. js," *Node: Up and Running* (1st Ed.), O'Reilly Media, Sebastropol, CA, pp. 4–11, Apr. 1, 2012.

133. J.R. Wilson, "Node. js the right way," *Pragmatic Programmers*, 2014.

134. E. Allen, R. Cartwright, and B. Stoler, "DrJava: A lightweight pedagogic environment for Java," In *Proceedings 33rd SIGCSE International Conf*erence, vol. 34, no. 1, pp. 137–141, Feb. 27–29, 2002.

135. P. Teixeira, "Professional node.js: Building javascript based scalable software," John Wiley & Sons, Hoboken, NJ, Oct. 1, 2012.

136. P. Tatade, "Why node, the fundamental difference between Node and other languages," Online, www.cuelogic.com/blog/why-node-the-fundamental-difference-between-node-and-other-languages, Feb. 2014.

137. I. Fette, "The websocket protocol," *Internet Engineering Task Force (IETF)*, pp. 5–6, 2011.

138. V. Pimentel, and B.G. Nickerson, "Communicating and displaying real-time data with WebSocket," *IEEE Internet Computing*, vol. 16, no. 4, pp. 45–53, Jul. 2012.

139. V. Wang, F. Salim, and P. Moskovits, "The definitive guide to HTML5 WebSocket," Apress, Berkeley, CA, vol. 1, Mar. 21, 2013.

140. Y. Furukawa, "Web-based control application using WebSocket," In *Proceedings ICALEPCS International Conference*, Grenoble, pp. 673–675, Jan. 2011.

141. Q. Liu, and X. Sun, "Research of web real-time communication based on WebSocket," *IJCNSS*, vol. 5, no. 12, pp. 797–801, Dec. 14, 2012.

142. R. Rai, "The Socket.IO protocol," *Socket.IO Real-time Web Application Development, Sebastropol: O'Reilly Media*, Packt Publishing Ltd, Birmingham, pp. 87–91, Feb. 2013.

143. M. Armbrust, I. Stoica, M. Zaharia, A. Fox, R. Griffith, A.D. Joseph, R. Katz, A. Konwinski, G. Lee, D. Patterson, and A. Rabkin, "A view of cloud computing," *C Communications of the ACM*, vol. 53, no. 4, pp. 50–58, Apr. 1, 2010.

144. L. Youseff, M. Butrico, and D. Da Silva, "Toward a unified ontology of cloud computing," In *Grid Computing Environment Work GCE 08 IEEE*, Austin, pp. 1–10, Nov. 12, 2008.

145. R. Buyya, R. Buyya, C.S. Yeo, C.S. Yeo, S. Venugopal, S. Venugopal, J. Broberg, J. Broberg, I. Brandic, and I. Brandic, "Cloud computing and emerging IT platforms: Vision, hype, and reality for delivering computing as the 5th utility," *Future Generation Computing Systems*, vol. 25, no. 6, pp. 599–616, Jun. 30, 2009.

146. S. Seiler, "Current trends in remote and virtual lab engineering. Where are we in 2013?," *International Journal of Online Engineering (iJOE)*, vol. 9, no. 6, pp. 12–16, Nov. 3, 2013.

147. P. Mell, and T. Grance, "The NIST definition of cloud computing recommendations of the national institute of standards and technology," *NIST Special Publication*, vol. 145, no. 1, p. 7, Jan. 2011.

148. A.T. Velte, and T.J. Velte, "Cloud computing : A practical approach," McGraw-Hill, New York, pp. 1–55, Jan. 2009.

149. N. Fernando, S.W. Loke, and W. Rahayu, "Mobile cloud computing: A survey," *Future Generation Computer Systems*, vol. 29, no. 1, pp. 84–106, Jan. 31, 2013.

150. P. Jamshidi, A. Ahmad, and C. Pahl, "Cloud migration research: A systematic review," *IEEE Transactions on Cloud Computing*, vol. 1, no. 2, pp. 142–157, Jul. 2013.

151. N.K. Jangid, "Real time cloud computing, data management," *Secure*, www.greenrayitsolutions.com/publications/rtcc.pdf, Jan. 1, 2011.

152. K. Goldberg, and B. Kehoe, "Cloud robotics and automation: A survey of related work," *Technical Report UCB/EECS-2013-5*, pp. 1–12, Jan. 27, 2013.

153. A. Biharisingh, "Build your first mobile app with the ionic framework – Part," http://gonehybrid.com/build-your-first-mobile-app-with-the-ionic-framework-part-1, posted, Jan. 23, 2015.

154. P. Orduña, J. García-Zubia, L. Rodriguez-Gil, J., Irurzun, D. López-de-Ipiña, and F. Gazzola, "Using LabVIEW remote panel in remote laboratories: Advantages and disadvantages," In *Global Engineering Education Conference (EDUCON)*, IEEE, pp. 1–7, Apr. 17–19, 2012.

155. F. Esquembre, "Easy Java simulations: A software tool to create scientific simulations in Java," *Computer Physics Communications*, vol. 156, no. 2, pp. 199–204, Jan. 1, 2004.

156. S. Gerkšič, G. Dolanc, D. Vrančić, J. Kocijan, S. Strmčnik, S. Blažič, I. Škrjanc, Z. Marinšek, M. Božiček, A. Stathaki, and R. King, "Advanced control algorithms embedded in a programmable logic controller," *Control Engineering Practice*, vol. 14, no. 8, pp. 935–948, Aug. 31, 2006.

157. G. Valencia-Palomo, and J.A. Rossiter, "Programmable logic controller implementation of an auto-tuned predictive control based on minimal plant information," *ISA Transactions*, vol. 50, no. 1, pp. 92–100, Jan. 31, 2011.

158. S.L. Herman, and B.L. Sparkman, "Electricity and controls for HVAC/R," 6th ed. Cengage Learning, Delmar, pp. 531–600, 2010.

159. J.A. Rehg, and G.J. Sartori, *Programmable Logic Controllers*, 2nd ed., Prentice Hall, Upper Saddle River, NJ; London, pp. 1–99, 2009. ISBN-13: 978-0135048818

160. F. Ley, "Programmable logic controllers—Architecture and applications," *Gilles Michel Automatica*, vol. 28, no. 3: pp. 652–653, Sep. 27, 1992.

161. Z. Aydogmus, and O. Aydogmus, "A web-based remote access laboratory using SCADA," *IEEE Transactions on Education*, vol. 52, no. 1, pp. 126–132, Feb. 2009.

162. R. Marques, J. Rocha, S. Rafael, and J.F. Martins, "Design and implementation of a reconfigurable remote laboratory, using oscilloscope/PLC network for WWW access," *IEEE Transactions on Industrial Electronics*, vol. 55, no. 6, pp. 2425–2432, Jun. 2008.

163. G. Farias, R. De-Keyser, S. Dormido, and F. Esquembre, "Developing networked control labs: A Matlab and easy java simulations approach," *IEEE Transactions on Industrial Electronics*, vol. 57, no. 10, pp. 3266–3275, Oct. 2010.

164. E. Fabregas, G. Farias, S. Dormido-Canto, S. Dormido, and F. Esquembre, "Developing a remote laboratory for engineering education," *Computers & Education*, vol. 57, no. 2, pp. 1686–1697, Sep. 30, 2011.

165. Z. Magyar, and K. Žáková, "Scilab based remote control of experiments," *IFAC Proceedings Volumes*, vol. 45, no. 11, pp. 206–211, Jan. 1, 2012.

166. A. Turan, S. Bogosyan, and M. Gokasan, "Development of a client-server communication method for Matlab/Simulink based remote robotics experiments," In *Processing of the 2006 IEEE International Symposium on Industrial Electronics*, Montreal, vol. 4, pp. 3201–3206, May 8–11, 2006.

167. L. Ciortea, C. Zamfir, S. Bucur, V. Chipounov, and G. Candea, "Cloud9: A software testing service," *ACM SIGOPS Operating Systems Review*, vol. 43, no. 4, pp. 5–10, Jan. 27, 2010.

168. R. Chamberlain, and J. Schommer, "Using Docker to support reproducible research," http://dx.doi.org/10.6084/m9. figshare, 1101910, Jul. 14, 2014.

169. J. Fink, "Docker: A software as a service, operating system-level virtualization framework," *Code4Lib Journal*, vol. 25, p. 29 Jul. 21, 2014.

170. J.J. Kuffner, "Cloud-enabled robots," In *Processing of the IEEE-RAS International Conference on Humanoid Robotics*, Nashville, TN, Nov. 15–17, 2010.

171. Q. Hu, and E. Johnston, "Using a wiki-based course design to create a student-centered learning environment: Strategies and lessons," *Journal of Public Affairs Education*, pp. 493–512, Jul. 1, 2012.

172. R.S. Brower, and W.E. Klay, "Distance learning: Some fundamental questions for public affairs education," *Journal of Public Affairs Education*, pp. 215–231, Oct. 1, 2000.

173. C.P. Coutinho, and J.B. Bottentuit Junior, "Collaborative learning using wiki: A pilot study with master students in educational technology in Portugal," In *Processing of the World Conference on Education Multimedia, Hypermedia and Telecommunications*, Vancouver, Canada, pp. 1786–1791, Jun. 25–29, 2007.

174. B. Schroeder, and G. Gibson, "A large-scale study of failures in high-performance computing systems," *IEEE Transactions on Dependable and Secure Computing*, vol. 7, no. 4, pp. 337–350, Oct. 2010.

175. B. Yalvac, M.C. Ayar, and F. Soylu, "Teaching Engineering with Wikis," *International Journal of Engineering Education*, vol. 28, no. 3, pp. 701, Jan. 1, 2012.

176. S.R. Tsai, Y.S. Huang, U.L. Dai, G.H. Huang, C.M. Lee, J.Y. Fang, Y.T. Chen, and J.N. Lee, "An object-based web system for building a virtual community," In *Processing of the 2010 Sixth International Conference on Intelligent Information Hiding and Multimedia Signal Processing (IIH-MSP)*, Darmstadt, pp. 414–417, Oct. 15–18, 2010.

177. M. Krötzsch, D. Vrandečić, and M Völkel, "Semantic mediawiki," In *Processing of the Semantic Web-ISWC 2006*, Springer Berlin Heidelberg, vol. 4273, pp. 925–942, Dec. 1, 2006.

178. K. Ogata, "Modern control engineering," 5th ed., Prentice Hall, Upper Saddle River, NJ Jan. 1, 2009.

179. P.M. Meshram, and R.G. Kanojiya, "Tuning of PID controller using Ziegler-Nichols method for speed control of DC motor," In *Processing of the 2012 International Conference on Advances in Engineering, Science and Management (ICAESM)*, Nagapattinam, pp. 117–122, Mar. 30–Apr. 1, 2012.

INDEX